AMANDA JUST

The Advantage Code

Equip your child to thrive in the age of AI

Dedication: I dedicate this book to my two beautiful daughters who have been my corner stone throughout this process. Their patience and consistent help has been instrumental in producing this book which I am immensely proud of. I would also like to thank the many people who have encouraged me to keep going and publish. Without you all, this project would not have continued.

"You only need one good teacher in your life to set you on the right path - and for our family that teacher is Amanda"

Contents

Preface

When I look back on my journey as both an educator and a parent, one theme keeps resurfacing: curiosity. Curiosity about how children learn, curiosity about how technology reshapes our world, and curiosity about how we, parents and educators alike, can guide the next generation with both confidence and compassion.

My path has taken many turns. I began at university doing a Bachelor of Science and Information Technology and on completion started my career in the IT (Tech) industry, immersed in the world of innovation and problem-solving. Later, I found myself teaching violin, watching children's faces light up as they unlocked the joy of music. Along the way, I pursued a Master's in Education to deepen my understanding of how children think, grow, and thrive. That same curiosity also led me to create a game that helps children learn to read music, a small window into how play and creativity can make learning magical. I continue to enjoy teaching children and watching them thrive to this day.

Parenting, of course, has been my most humbling and rewarding role of all. My two beautiful daughters have been my greatest teachers, constantly reminding me of the challenges, surprises, and laughter that come with raising children. Together, we even launched Modern Successful Parenting, a website and podcast where we spoke candidly about the real issues families face today. More recently, I've released free parenting advice clips on YouTube, because I believe every parent deserves access to support and encouragement, no matter where they are.

Over the past 2 decades, I've had the privilege of sharing these insights in parenting lectures across Australia, the United States, and Japan, facilitating workshops, and helping families navigate the challenges of raising resilient, future-ready children. Each conversation reaffirmed something I believe deeply: no matter where we come from, parents share the same hopes and worries. We want our children to be happy, resilient, and prepared for a future that feels increasingly uncertain.

This book is my way of bringing those experiences together. It's not a manual with rigid rules, but rather a collection of insights, stories, and strategies gathered from years of listening, teaching, learning, and parenting. My hope is that you will find reassurance in its pages, that you'll feel less alone in the challenges, more confident in your choices, and more excited about the possibilities that lie ahead for your children, you don't need to have all the answers, you just need to walk alongside your child with openness, courage, and a willingness to grow together.

I invite you to read with curiosity, to take what resonates, and to adapt the ideas here to the unique needs of your family.

AMANDA'S UNIQUE PERSPECTIVE

INFORMATION TECHNOLOGY:

Bachelor of Science & Computing
Integration specialist
Business analyst
Web designer
Small business owner

PARENTING:

Over 20 years parenting
Parenting lectures worldwide
Parenting advice 20+ years
My own children

EDUCATION:

Masters of Education
Suzuki Violin
Grad. Diploma of Primary
Music Specialist Teacher
Technology Specialist Teacher

1

FUTURE-READY CHILDREN FOR THE NEW WORLD OF AI

Little did I know when my Dad bought home the latest Apple IIe into our house, what an impact this would have on my life and that of society as a whole. This unassuming piece of equipment was fascinating to me. He insisted that his 'little girl' learnt typing in year 9 and 10 even though her focus was science. He believed that typing was going to be an important skill in the new world evolving before his eyes, even though most of the girls learning it were focused on a different employment pathway. He was right. Typing became the most valuable skill I learned in middle school. Not because it directly landed me a job, but because it has shaped nearly every aspect of my life. At university, I typed assignments for others to earn money improving my own writing skills in the process. Later, it gave me opportunities to teach CEOs how to use computers, building relationships with influential leaders. Typing wasn't just a skill; it was my gateway into technology and professional growth.

How could my Dad have known how valuable a skill as simple as typing would be? I remember the distain which people had while commenting to him "Why would you get her learn that, she is an academic kid?". Well my Father knew something most other people didn't, typing was an essential skill for the new world of computers. He was going to skill up his little girl for the new world. Today, the "Apple IIe moment" for our children is AI.

So what is Artificial Intelligence, commonly known as AI? It is basically a computer program which learns like the human brain. It processes enormous amounts of data, detects patterns, and makes predictions or recommendations. It is a new rapidly developing technology which is quickly spreading world wide, showing us we may need to be more careful than we think. One issue we will face with AI that stands out to me, is the validity of how AI is programmed to learn. While I'm writing this book there are new laws being written in Europe trying to ensure the safety of the knowledge which AI is producing. It is a new developing world in which AI is able to do many tasks and jobs us humans do currently. We will need to consider many factors when approaching working life as jobs change or even become non existent. While we adjust to this new technology how can we also ensure that our children are ready for the future?

"The human connection is what stands us apart from AI"

The answer lies in what sets us apart from machines: human connection, creativity, values, and wisdom. AI may process data, but it doesn't truly *care*. It doesn't understand human emotions or live by values. What we teach our children, how they treat others, how they think critically, and how they protect themselves is what will matter more than ever.

Family values vary widely across cultures, beliefs, and circumstances. Some families prioritise charity and service, while others focus on survival. Yet no matter where we begin or how our past looks, what we pass on to our children becomes their greatest advantage in an AI-driven future. This may look like our own habits, problem-solving techniques, or displays of our resilience. AI is leveling the playing field and asking us to re-evaluate what will be important for success in the near and far future. It's not just about knowledge anymore, it's more than ever about how we interact with others, the world around us and technology.

Human Centred Skills

The most important concepts which our children need to understand to succeed are our human-centred skills that AI cannot replace. The most important are critical thinking, creativity, caring and how to protect their data;

• **Critical thinking** is essential to help us analyse AI output, create new innovative ideas and question the world around us. AI can only take what has been done and extrapolate. There is no innate ability to create new innovative thinking, leading to science discovery or research opportunities. This still needs the organic brain.

• **Creativity** is used in all aspects of learning and the arts. It's not just about creating new pieces of music or art, but also new robots, new ways of thinking or new ways to enhancing others lives.

• **Caring** is something which AI currently has no capability. It may come close in replicating caring speech but it will always be different to what comes from a physical person. Empathy, kindness, and compassion are uniquely human. These qualities build trust in relationships which are skills no algorithm can authentically replicate and are needed in todays society.

• **Data protection** is another area which is complicated in AI. children must learn how to safeguard personal data, protect intellectual property, and manage their online reputation from an early age. In an AI-driven world, protecting *who you are* is just as critical as protecting *what you know*.

FUN ACTIVITIES FOR HUMAN CENTERED SKILLS

CORE CONCEPT	SIMPLE EVERYDAY ACTIVITIES	WHY IT MATTERS
Critical Thinking	At the dinner table ask: "What do you think about...?" or "Why do you think that happened?" Encourage explanations, not just answers.	Helps children analyse AI outputs and spot when something doesn't make sense.
Creativity	Keep a "what if" journal: "What if animals could talk?" or "What if we could live on Mars?" Brainstorm ideas together.	Creativity fuels innovation — the one area AI cannot replicate.
Caring	Model empathy by narrating your thoughts: "I can see you're upset — let's work out what you need right now."	Emotional intelligence and compassion build trust — essential in human relationships and jobs.
Protection (Data & Reputation)	When your child asks for a photo online, ask: "Would you want future employers to see this?" Teach them about safe sharing.	Protecting personal data and reputation is critical in a digital world where mistakes live forever.

In essence it is more important than ever for children to learn these human centred skills. They are key to ensuring that your children are ready for a successful future. We can't deny that AI is here to stay.

In the future, people will still need caring for, people will still need to stay connected to others, people will still need essential services, however using AI to enhance these jobs is the key to success in the future.

In the following chapters we will go through how to prepare your child for the future. We have broken down what types of skills your child will need to master throughout their development to take advantage of AI in the new world. You will find that they are divided into three levels. You might look at these skills and wonder why they are essential for the AI world, but be assured we will take you through why, how and when for each one as well as giving practical tips and games to help you have fun with your child. Don't despair if you find that you have not introduced these skills in the time frame we suggest. Today is a good day to start to give your children The Advantage Code for future success in this new AI world.

2

A ROADMAP: BUILDING FUTURE-READY SKILLS IN AN AI WORLD

I have to admit that AI is changing the world faster than anything else I personally have ever experienced or read about in history. New industries are emerging, money is being made and we are seeing people winning and losing at AI. We need to move quick and respond to the change but our children are still small, their brains are developing and they need to experience the world around them in terms that they can understand. It's important that we don't lose sight of this as we guide them into the new AI world.

Raising children in the age of AI means preparing them with more than just academics. They'll need emotional strength, curiosity, adaptability, and practical skills to thrive in a world where technology is constantly evolving. That's why this book is structured around three levels of learning; **Foundation, Intermediate, and Advanced Skills.** Each skill builds on the other, yet is flexible enough to adapt to your child's pace and personality.

Here's how the journey unfolds:

1. Foundation Skills

These are the **early building blocks**: emotional intelligence, resilience, adaptability, and collaboration. They shape how your child understands themselves and others, and they lay the groundwork for every other skill. Think of them as the roots of a tree which are strong, deep, and essential for growth.

2. Intermediate Skills

Once the roots are strong, we move into the **branches of learning**. Here, your child learns to protect and verify data, build general knowledge, explore research, understand how things work, and practice problem solving. These skills help them connect the dots between curiosity and real-world understanding, so they can make sense of the flood of information AI brings.

3. Advanced Skill

Finally, we reach the **leaves and fruit.**These are skills that help your child thrive in the wider world and can be very useful in early adulthood. These include prompting AI effectively, marketing themselves with confidence, and mastering basic tech skills. Advanced skills integrate everything learned before and empower your child to lead, innovate, and stand out in the AI-driven future.

Throughout all three levels, you'll find:

- **Personal stories** to show these ideas in action.
- **Practical tips** you can use straight away.
- **Fun game tables** at the end of each chapter to make learning natural and enjoyable.

Remember: this isn't a race. Your child doesn't need to master every skill at once. You are also learning at your own pace, so start with what feels right for your family, and return to the others when the time is right. Each small step builds confidence, consistency, and capability.

Gradual Implementation of skills:

You will be relieved to hear that these skills I share are not anticipated to be implemented all at once either. I remember when I read a parenting book which told me how to deal with biting. There were 3 stages. First say 'No', then if they did it again, slap their hand and finally if they bit again put them in timeout in their room. When my daughter bit me, I panicked and did all three at once. I sat outside her silent room and thought "What now?". Needless to say she and I were so shocked that she never bit me again, and I learnt that I should really take more notice of what I was reading. It definitely was an overkill but we all make mistakes and she was fine.

Together, these three stages create a roadmap for raising future-ready children. Nurturing children who can think critically, adapt quickly, and use AI as a tool to create a life filled with curiosity, resilience, and have 'The Advantage Code' for success. Think of this chapter as a roadmap, an overview of what's to come.

♀ Fun Activities

The fun activities we have listed at the end of each explanatory chapter are a great fun way of re-enforcing the skills children need in an AI world. You will notice they are are separated into the three age groups.You may already be practicing some of the advanced skills without realising! If not, then it is never to late to start to get the extra benefits. Often activities can work for multiple skills at once and you will see them replicated across the tables provided.

▨ FOUNDATION SKILLS OVERVIEW

Suggestion: Start at 2 years

The foundations skills are just that, the skills which will sit at our base and allow us to build on top. Luckily most of us can build our foundations on our own, but if we are given a bit of help our foundations can be built with extra strength. Who doesn't want to give their children the foundation which can hold the 2 story house instead of the 1 story? Not only that, but for children and teenagers there really is no time their foundation skills cannot be built up or improved. That's the beauty of foundation skills: you can begin them early (often from age one or two) and then layer, practice, and refine them throughout childhood and adolescence. Even teenagers benefit just as much as toddlers from knowing these basics (even I benefit from a refresher every now and then).

❤ Emotional Intelligence

Emotional Intelligence (EQ) is the ability to understand, use, and manage emotions. Adding these skills, with a healthy ability to empathise, allows us to also communicate effectively with others. A high emotional intelligence helps us to manage stress, overcome challenges in life and even contributes to academic success. It helps with understanding the people around us and what is going on in our own heads. Basic Emotional Intelligence includes skills such as understanding your emotions and controlling your emotional responses. This is essential for children so they can succeed in building strong relationships and over come adversity. They gain control over their own emotional responses and can succeed in the workplace, social life and beyond.

Everyone learns at different rates. To some, it comes easy, however others need to be taught and guided in ways to control their responses to the emotions. Notice how I said 'control their responses to the emotions'. It is important that children understand it is ok to feel emotions, it's your response and actions which we have control over and can change. That's the parents job, to guide and teach children in how they respond to their own and others emotions, giving them the edge in the future job market.

Social and Living Skills

Social skills and living skills seem like second nature to us adults, but children will need time and repetition to learn them. I sometimes don't want to 'adult' and wish I was a kid again, however it is important that we teach our children as much as we can about how our society works and how to 'adult'. It is all about helping them to become successful and believe me when I say that there is a lot involved.

We start by teaching them independence in their everyday lives, such as, getting dressed, cleaning their teeth and putting shoes on. I remember standing and waiting until my 2 year old got everything ready for a trip to the shops. I would have to make sure I left at least 30 minutes for the whole process. It was very frustrating when I had deadlines or appointments to get to, but I knew it was important for her to learn how to do this herself. Luckily she did get quicker over time but I would always give a "We are leaving in 5 minutes" to hurry her along. This is the way little ones brains grow and develop. Through *repetition*.

The question remains, why do they need this type of experience in an AI world? Actually, having experience in how a household works, will give your children the basic data they need to direct AI in their lives. I love my robot vacuum which I can direct to clean up any room at anytime. If I didn't understand why it was so important to have a clean house, I wouldn't know to start that little helper. Teaching your child how to be independent using basic house skills, cleaning, and cooking, will give them the data they need to make their lives easier in the future. Not only does it teach them how to live healthy lives, but they get to physically experience the world developing their living skills. Children need the physical experiences from life, as this is how their brains organically develop and grow.

♖ Debating Skills

Being able to debate an idea is an important skill for many aspects of the new AI world. Firstly, you can use these debating skills to ensure that you are heard in the work place. Formulating your arguments of why the business should listen to your ideas can give you a head start on your career pathway. Secondly, AI systems train on large datasets, that contain biases, inequalities and historical inaccuracies. In short, they often use popular opinion instead of actual fact (readily available data is not always accurate data). If you are trained in debating skills you can identify and critically analyse the AI output for any inconsistencies and biases.

But beware, better debating skill can come back to bite you. When my children were younger we went on long 8+ hour drives to go on holiday across states. We often would start a debate in the car to pass the time. The practice allowed the children to learn the process of debating and how to formulate an argument. But after some time I wasn't sure if I created two little monsters when they would come up with innovative reasons why they should have a certain toy. The reasons were so good I would find myself giving in and buying the toy. This was a good lesson for them, and one for me. Don't start a debate you aren't sure of the outcome, otherwise you could end up with a new puppy.

✍ Collaboration Skills

Humans are wired for community. From family dinners to workplace projects, collaboration forms our ability to thrive together. Collaboration is one of the keys to community and is even taught as a foundation skill in schools today. You can start learning these collaborative skills as early as 2 when your family (a team) works together to clean the house, or cook a meal. This is the corner stone of collaboration and their first taste into how wonderful it feels to achieve an outcome together especially as a family. In essence, humans are a community society who works best surrounded by family and friends. It is much more fun doing activities with your friends and its also scientifically proven that you are much less likely to have mental health issues when you have a strong support system. Our house was always open to anyone. People would come and go. We would often have other children staying for dinner without a second thought to help out our friends in need. It was wonderful having such a busy household. The earlier in life children learn to collaborate, the richer their lives will be.

Collaboration is usually observed when we come together in teams and combine others critical thinking, creativity and judgement. Adding collaboration alongside AI will not only achieve better outcomes in strategic decision-making but possibly more innovative solutions. AI is for data analysis but people bring wisdom and world experience to the AI-generated outcomes. If we can teach our children to use their own collaboration skills, alongside AI, we will create a new better world with innovative solutions to some of our long-term problems. It's all about learning to work with others and to get the best outcomes.

🏛 Resilience

Resilience is the ability to recover from a bad situation. Any type of misfortune (challenge or difficult life event) can cause stress and low/high emotions. Being resilient means that you are able to reduce this stress, have a great outlook on the situation, work through the issues and come up with a plan of action to fix the problem. Sometimes it can be as simple as knowing when to walk away. It's about being confident in your own abilities to cope in difficult situations and with setbacks.

Showing your children that resilience is important can be easily done through creating small resistance in their own lives and praising them when they learn how to glide through the bumps. By that I mean that you can say 'No' to a new toy which everyone else has. Guide them through the emotions and what actions they should do. After a few times, they will learn to just shrug it off. By praising them you can re-inforce the behaviour and voila, you have resilient children. Resilience is so important as AI is already mounting challenges which we cannot predict, so having resilience helps our children to be able to cope with the continuous changes and fast pace. Being resilient will give our children the edge they need to be able to work with AI in the new world.

⌂ Grit

Grit is a term created by Angela Duckworth. She uses this term to describe the ability to push through the hard part of learning, to end up with a new skill. These skills are all a result of the learning process but will require the Grit element to really finish the lesson. Grit can refer to a physical skill, like learning a new piece of music, or even just learning how to write your name for the first time. It can also refer to an academic skill, like learning how to read or do math as well as an emotional skill like learning how to lose gracefully.

There are going to be many times in an AI world where our children are going to be faced with new learning opportunities. Some of which we have never even thought about yet. They need to be able to understand how repetition is part of learning, as well as how they can manage the disappointment they may face. They need to understand that when they get it wrong, it is temporary and through repetition they will attain the skill. With Grit they know not to give up part way through but to keep going when it's hard. Children will understand that they can learn anything with Grit. It is worth checking out Angela Duckworth's books and Ted talks (see the end of this book for where to find these).

Adaptability

Adaptability is not just about "going with the flow" and knowing how to "deal with change". It is also about how we are able to **change** our thinking and processes based on the information we receive. children need structure to cope in their world. If you think about it, children change on a daily basis. Their bodies grow, their minds develop and their hormones change. They are in essence dealing with change everyday. Adaptability in a child, means that they can thrive when routines change, the environment changes or the people around them change. It's about coping when things go wrong, shifting strategies and accepting mistakes or failure. AI is changing our lives and how the world works, so being adaptable will help your children to be successful in the future.

The Bottom Line

Foundation skills aren't quick lessons; they're lifelong habits built through repetition, modelling, and everyday experience. Each one helps children grow into capable, resilient adults ready to thrive alongside AI.

✿ INTERMEDIATE SKILLS OVERVIEW

Suggestion: Start from 5 years

Intermediate skills are generally taught from the age of 5 - 8+ years. They are more complex as the children's brains are further developed and becoming aware of the world around them and how they fit in. Children from this age have practiced, but not mastered the foundation skills. Though learning the foundations will be an ongoing process, they will now know enough at this point to integrate these new skills into their repertoire.

▦ Value of Data - Critically Assessing Information

AI is only as good as its data. AI takes the most popular answers and feeds this to you unless you specifically ask it to source data from specific areas. It's quite dangerous as the most popular answer may be far from the truth. It's like the idea that the Great Wall of China is the only man-made structure visible from space. People have been quoting this for years whereas in reality, it is not even the most visible structure from low Earth orbit. This popular belief would be presented as truth if you asked AI without narrowing the prompt down.

Children need to understand that the more relevant the data, the better the outcome. They need to be able to pick what is good data and how can we get the most information out of it. Teaching our children how to analyse and critically assess the outcome as well as where the data comes from helps them to interact with AI in an educated way. The key to being able to assess this information is general knowledge.

They need to know how to analyse all new information within the framework of what they know about the world. Sometimes this might mean that they need to adapt and change their thinking based on the new information.

Working with AI instead of just using AI can really help in this skill. Knowing what questions to ask, or what to learn more about will give them the edge.

💬 General Knowledge

General knowledge is simply knowing a little about a lot. It comes from books, stories, conversations, travel, and hands-on experiences. Sometimes it is about historical events, science or nature facts, how something is put together or even manufactured. Actually it is about anything and everything in our world. My daughter use to say that she was the gate post - silent but listening all the time. She built up a great general knowledge about how the world works through listening to the conversations I was having with other adults as well as watching and participating in DIY projects. She was fascinated with trades and loved watching how plumbers fixed drains and taps when they broke. The other day when one of her friends backed into the garden tap at her house, she jumped into action, turning off the water to the house, going to Bunnings and getting the items needed and fixing it herself. Skills such as critical thinking, problem solving, collaboration and strategic thought processes, require a good general knowledge, as does academic learning, innovation and creativity. General knowledge is so important in all parts of life.

The key to learning general knowledge is the memory, the ability to learn and retain facts. To relate this to technology and specifically AI, if you have enough information in your head about the world in general then you can easily work out if AI is telling the truth. If it doesn't seem like the truth AI may be using the unsubstantiated data. Without the ability to notice what is fact from the information output ourselves or our children cannot critically assess and use the information which AI provides.

⚖ Knowing How Things Work

Having a healthy general knowledge is only as good as knowing the context that goes along with it. In essence without knowing more about the system the fact relates to, you have very limited use of the fact itself. As a real world example I recently put on a clip for my class at school about how packet chips (or crips) were made. Some children had no idea that chips are made from potatoes. To see their faces squish when all the dirt is coming off the potatoes and show surprise to see how many slices were made from each potato was priceless. Even seeing how they were packaged was a new discovery. They now knew that chips were made from potatoes but also understood that potatoes come from the garden and the chips become chips from preparation and packaging. The extra context of how this process worked gave them much more useful knowledge than the simple statement "chips are made from potatoes".

My class then started to ask questions about other packaged food. They started inquiring and searching for answers. How are lollies made? What came before the lollies? How is sugar grown? This love of learning is what we need in the AI world. The love of learning will lead to developing their general knowledge and knowledge of how things work. In combination this gives children an edge in AI as it will help them to understand if what AI is saying is correct. In an AI world, it's not about remembering the exact technical detail but knowing the general gist and what it achieves.

📚 How to Research - an Inquiring Mind

Research is more than typing a question into Google (or asking AI). It's about knowing how to ask good questions, compare different sources, and decide what's reliable. Being able to research is a skill which is developed over years and should be guided by our teachers during school. Learning to ask the right questions, find the right resources and come to the inevitable conclusions and answers. These skills are used in everyday life but can now be applied when using AI, in the form of a good prompt.

We know that AI can do the heavy lifting of research, however it is limited to data analysis only and doesn't form innovative ideas. Children can use AI to get the data they need to support their research and even help to write the thesis, but the questions to start with, will come from the human brain ingenuity.

You can start teaching your children how to research at quite an early age, even as early as age 4. There are great games and ideas in the following chapters to practice these skills. Remember that the hidden gem of showing our children how to research ideas is that they will develop a love of learning for life.

💡 Problem Solving Skills

Problem solving is an essential skill which is used every day for almost everything we do. Children watch and learn how to solve problems constantly, whether by watching an adult fix a problem or through their own trial and error. It can be as simple as working out what to do if you don't have plates on the table ready for dinner, or finding your shoes because you didn't put them in the right place. All of these small problems are a form of problem solving. The more little problems they learn how to solve, the more skills they have in solving the bigger problems, be it academic or life. Allowing them to problem solve themselves can build their confidence helping them attempt to solve bigger problems on their own more often. Nurturing and developing our children's problem solving skills is an important job for us. It can be developed at school, but we can start it from an early age at home.

Comprehending the problem is the first step in solving the problem, so be a Mentor to them. Remember that you are their guide in life, so you will sometimes have to count to 10 before you jump in and tell them a solution. You might be surprised with their solutions. You also might even have to let them fail in order for them learn how to problem solve solutions. Small failures helps them to learn the skills needed to keep going (grit) and perseverance (resilience) to come up with a solution.

In an AI world we are becoming reliant on AI to provide solutions to all of our problems, however using our own brains to critically think about the issues and come up with alternative solutions will put our children ahead of the majority of job seekers in the future. Ensuring that they have practice in finding lots of ways to solve the issue is essential for their development and growth to maturity.

The Bottom Line

Intermediate Skills build on the Foundations Skills by encouraging curiosity, critical thinking, and independent exploration. They help children question data, seek credible answers, and use their growing knowledge of the world to problem-solve effectively.

In the AI era, these skills will make the difference between children who simply **use** technology and those who can **lead, innovate, and adapt with it.**

✬ ADVANCED SKILLS OVERVIEW

Suggestion: Start from 10 years, teens and adults

Advanced skills become most relevant as children move into their pre-teen and teenage years (around 10 years and up). Some of the basics of our Advanced Skills may already be in practice after working through the Foundation and Intermediate Skill repertoire. Think of these Advanced Skills as **fine-tuning**. They help older children connect their knowledge, curiosity, and resilience with the tools of technology and the demands of the AI era. At this stage, we're not just preparing them to use AI, we're preparing them to **work alongside it, guide it, and lead with uniquely human strengths**.

The fun activities we have listed at the end of each explanatory chapter are a great fun way of re-enforcing the skills children need in an AI world. You may already be practicing some of the advanced skills without realising! If not, then it is never to late to start to get the extra benefits.

Prompting Skills and Critical Thinking

AI's usefulness depends entirely on how we interact with it. A poorly written question (or "prompt") will usually give a vague or even irrelevant answer. You need to be specific and more importantly, you need to know what you want before you can ask for it. This is an obvious statement but something which is core to AI and knowledge engineering software. This is where our ability to be able to prompt and analyse what AI gives us, is so important (applying our intermediate skills).

Prompting skills are quite simple and we only need enough knowledge to understand what it is we need. We don't need the actual detail. It's like the **percentage button %** on the calculator. We all know what a percentage is, and how you use it. But how many of you know how to calculate it manually? Since our new calculator technology we just press the button and it comes up. We don't need to know the formula, just know what the concept is, and how to use it. This is exactly like AI. We need to know what the right question is and how to analyse the result. Once this is done the second step is to determine if the answer provided was even appropriate. This is where the critical thinking, general knowledge and how things work skills apply. Making sure you can actively ask questions and evaluate information, ideas and beliefs.

Developing these types of skills in our children is as easy as talking through how you are thinking when things happen in your lives. We have all the content in our own lives to help children with these skills.All we need to do is let our children know what we are thinking.

✍ Marketing Skills and Strategies

Now that the Foundation and Intermediate Skills are in our children's repertoire we can teach them some skills which really become important in their late teen/adult years. In my experiences (as a teacher and parent) I've seen how marketing and strategising skills really impact how successful young adults are, especially for employment opportunities.

Finding work is becoming increasingly difficult and finding your place in the world once leaving school or entering the workforce is going to be harder when AI can do so much. It's all about how we use AI and what human qualities we bring to the job. Marketing skills will aid our children's ability to use creativity, communication, strategic thinking and data analysis to promote anything they wish (including themselves). Once we help them to add a marketing strategy to those skills, they will be able to develop their own plans to convince people they have the right solutions. It's about learning the process of marketing so you know how to market yourself to land that new job in the new AI world (skills we can all benefit from).

Just as we have discussed before, AI already has the ability to act like a specialist. It can even act as a marketing guru, so why do children need to be able to market things? Marketing skills and strategies can give them the ability to describe what they have to offer companies, show employers why they can add value to a job or even understanding market trends to arrive at the right solutions.

AI is not flexible enough yet to see outside the box and innovate. It can teach you and provide great marketing ideas but essentially teens and young adults need to know that marketing exists and how the process works so they can ask the right questions. So that they can USE AI for a good outcome. This is the era of the generalist. A jack of all trades. Understanding generally what is needed, but using AI to get the details.

💻 Practical Technical Skills

Technology is woven into nearly every aspect of modern life, and our children need a strong foundation in **practical tech use**. While I recommend limiting screen time in the early years, once children reach school age they'll need to begin mastering essential skills. Some of these skills include touch typing, saving documents, file organisation, and cyber security. Later it is important to talk about social media, search and verification skills as well as trouble shooting basic problems with tech devices.

Its important to note that only using a device *basically* is not enough to develop advanced practical technical skills. Though a child may have been using a device since their early years it doesn't necessarily mean they have been learning the skills to take advantage of all technology can provide. Sometimes their brains are too young to learn these skills and all they can manage is the basic actions of using programs. For this reason and many others we should all consider the positive and negative outcomes when giving children early screen access.

The Bottom Line

Advanced Skills aren't quick lessons; they're lifelong habits built through repetition, modeling, and everyday experience. You cannot necessarily expect them to be taught at school either. Mastering these skills will give your child the edge when competing for jobs. Each one of these help children grow into capable, resilient adults ready to thrive alongside AI.

3

FOUNDATION SKILLS IN DEPTH

Alright friends, strap in — we're about to dive into some big chapters packed with practical tools and important ideas. While many of these skills can be introduced from as early as age one, they aren't just for toddlers. They're **life-long lessons**, continuously built upon right through adolescence and into adulthood.

These **foundation skills** form the backbone of everything your child will need to thrive in a world shaped by AI. They support not only academic growth, but also emotional intelligence, resilience, adaptability, and the everyday problem-solving that future-ready children will rely on.

Inside each chapter, you'll find:

- **Tips and strategies** you can use straight away.
- **Stories and examples** to bring the ideas to life.
- **A game table at the end** of every chapter, giving you fun, age-appropriate ways to practice each skill at home.

My advice? Grab a highlighter and mark the tips that resonate most with your family. Start small, and build from there. Above all, remember this:

Consistency and repetition are the keys to success.

The more often a child practices a skill, the more it becomes second nature and something they can call on without even thinking.

Screen time - pros and cons

Before we go any further I would like to address the suggest time for screens for children under 5. It is 30 minutes or less. If you are already over that, don't despair you can reduce it little by little to get to the desired amount. The reason for this is the impact screens have on the developing brain. Radiology scans of children's brains who were using screens for more than 6 hours a day is frightening. Their brain matter is significantly different to other children who watched screens for less than 30 minutes. The research is conclusive. In a study conducted by Muppala, Vuppalapati and Pulliahgaru in 2023, they also concluded that although it is thought that screen media can have a positive impact on potential learning, it actually has a negative affect on children's executive functioning, sensorimotor development and academic outcomes in later years. This has a big impact for their future. It was also found that screen time can lead to impaired social-emotional development, sleep disorders and anxiety. The suggestions which have come out of these studies is that screens should be limited to children in the early years. They will have plenty of time to catch up once they are at school.

I often hear parents say that my children love the screen and are glued to it. Well yes, screens are addictive however because of the flight vs fight instinct we all have, they actually cannot look away. Let me explain.Screens flicker constantly. We have an instinct as humans that our sight is drawn to things that move. It is bred into us to protect us from danger (flight vs fight instinct) If something moves in our vision, even if it is on our peripheral vision, we look. If there is a sound we are drawn to see what it is. I bet you have heard a door slam and look in that direction, or see a car move into your vision and look at it straight away. This is a human instinct to make sure we are safe and can run away from danger. Screens constantly flicker so little people are drawn to the flicker and **can't look away** as the screen is constantly drawing their attention to it. As adults we know it is not dangerous so we can look away, but their growing developing brains are saying that it might be dangerous so they need to keep watching. I have been in very heated conversations on this topic, with people saying that children can start to learn reading at 2 years of age on the iPad so this is giving them the academic edge.Research has shown that waiting will give them superior skills in the future. They have plenty of time to catch up when their brains are developed a little more at school age.

So, let's get started on laying the foundations they need for the future AI world.

❤ EMOTIONAL INTELLIGENCE

Helping children to have a strong Emotional Intelligence (EQ or EI) is one of the most important skills you will teach your child. This is why I mention it as my first Foundation Skill.

The definition of emotional intelligence (EQ or EI) is the ability to understand, use, and manage your own emotions in positive ways to relieve stress, communicate effectively, empathise with others, and overcome challenges. As our children's brains grow and develop we see them gain an understanding of their emotions and how to cope with them. While they learn to cope with new emotions we have an opportunity to guide them differently through their EQ development.

As children's brains and bodies grow through each of the developmental stage they see the world in a whole new way. The release of hormones and life experience throughout this process helps them to understand the world around them in more detail. This definitely impacts their emotions and their ability to understand and control them. You might think that they are beautiful angel's one day and then the next day they wake up and seemingly have forgotten all the skills they knew and you are back to square one. This is very normal but quite frustrating for us as parents. We thought we had it sorted and then BOOM, new hormones, new emotions and another stage to go through.

I remember listening to my girls start fighting when they were teenagers. I thought all the fights were over until one day, one of them "borrowed" the other ones shirt. As she walked through the door with the shirt on, explosions erupted in my house. I couldn't believe it, it was just a shirt, but we all know that this was war in their minds. I ended up separating them until they cooled down and then the talks started with each one. Long long talks, but we ended up sorting it out. This is the essence of helping children through EQ and it continues into early adulthood.

While talking about EQ, it is important that we have a basic understanding of children's development. Piaget is a fundamental researcher of children's growth and development. His understanding of cognitive development has guided research for over the last 50 years and is still considered relevant today. Piaget broke up the development of children into 4 basic stages, Sensorimotor (0-2 years), Pre-operational (2-7 years), Concrete Operational (7-11 years) and finally Formal Operational (12+ years). A summary of his findings can be found at the end of the book. He deduced that children's brains learn in steps from learning about the world with their bodies and senses to understanding more complex hypothetical and abstract concepts when they are going through puberty. We also know that these steps are all influenced by how we feel and act on our feelings.

Emotions effect all areas of life and can eventually lead to success or failure in adulthood. Long term research studies have recently been completed looking at how children's emotional intelligence in the playground effects their success later in life. They were attempting to narrow down which skill is the most important factor for success (Gest et al, 2006). In this study it was discovered that kindness is that skill. Those happy, kind, popular children in the playground grew up to be the most successful adults. The study found that kindness, remembering truths about others, and caring for others (or appearing to care for others) were the top indicators of success in later years. I believe that these are also key skills for excellence in an AI world. Other studies which look at early indicators of success in adulthood also show that it is the human connection which makes people successful and apparently happier, but that's another book in itself.

The difference between AI and people is definitely the human connection. AI cannot currently provide kindness like humans can, neither can it show empathy or have emotional intelligence. Teaching our children how to control their emotions, look for emotions in others, read body language and how to build strong relationships with others, will give them the edge over others. Teaching our children a strong EQ, will also give them the insight of knowing when to respond to situations or just walk away. What a great gift you can give your child.

Let's look at Emotional Intelligence broken into the 3 age groups we are looking at.

Fundamental Skill Stage (1 - 5 years):

Children by one year of age are just becoming self-aware and have learnt to express their basic emotions. They should start to be aware of others feelings and even offer comfort to another child, adult or pet once they are coming up to the age of 5 years.

Mimicking (0-1 years)

I remember when my baby was giggling away and her father came in from the garden hurt. The change in her facial expression was immediate. She mimicked him exactly. Her tears started and I didn't know whether to help him or her. They were both just as upset. It really showed me how children are effected by those around them. A quick cuddle and reassurance fixed her tears and we could help daddy's hurt hand. In no time she was giggling again banging on her pots and pans.

I love watching a 1 year old mimic their parents during a conversation. It's amazing how accurate they portray the emotions which the parent is feeling. This mimicking helps them to start to understand the emotions they and others feel.

Object Permanence (0-2 years)

Along with this they are developing object permanence which means that they understand that things exist when they cannot see them.Sometimes it can be very upsetting for them when they know that Mum and Dad exist but aren't close. Often from 9 months to 1 year they can get a little separation anxiety which is completely normal. It is important that they work through these emotions. It might mean that they will cry for a bit, but that is ok as it will help them to understand and practise how to control their emotions.

This will help 10 fold in the future when things go wrong and are out of their control.

Defiance and tantrums (2+ years)

When they become toddlers their emotions become more complex as they start to experience emotions like shame, guilt and pride. This brings on the defiance and tantrums which I'm sure many have had to endure and are well aware of. Children start to be able to name their feelings and tell us what the problem is at 2 years but it doesn't make it easier on anyone when they are screaming at the top of their lungs. You will also see children at this stage starting to show concern for others and do pretend play about situations which work through some of the more complex emotions they are feeling.This **pretend play** is very important for their developing minds to sort out what is ok behaviour and how to work through the emotions they are feeling.

Empathy and Compassion (3-5 years)

From 3 -5 years they notice that others may not feel the same as them about a situation. This shows they are gaining new skills leading to empathy and compassion. Having friends to play with may become more important to them and they begin to want to please their friends and be liked by them. After all if your friend runs off crying the games has to stop. They can start to regulate their own emotions in situations which are familiar to them and have the ability to play co-operatively with others. Sometimes they find it difficult to know the difference between real and play/make-believe but they are very curious and start to become quite independent. This independence is very important for their self-worth and future.

Rules and Routines make life easier for everyone

HINT: To make your life easier it is important that children can anticipate what is going to happen. This way they have time to regulate their behaviour. Knowing routines and keeping rules simple will help them to understand the world around them and the feelings they are experiencing. If you sometimes say "It's ok to hit a pillow when you are angry" but then tell them off for hitting a person or breaking an object, it can be confusing. If they are confused, then there is another emotion they need to deal with. The situation compounds and we want to avoid this at all costs. My advice is to **keep the rules simple and consistent.**

Techniques to help with emotional regulation

The first step to emotional intelligence is to recognise and name their emotions. This can start when they are between 1 and 2 years. You may already be having lots and lots and lots of conversations about their emotions. This is perfect. The more you talk about emotions and help children to learn how to name their emotions, the easier it is later. Remember that it is the **repetition** of these conversations which will help the children to remember how to deal with the situation. As they get older we see them grow and start to control their tantrums, become more empathetic and co-operative.These skills develop over time but are the building blocks to a strong Emotional Intelligence.

Once they get to school they will use a system which is designed for use in schools. A lot of schools adopt the "Zones of Regulation" system which places a different colour for a group of emotions. This gives children a language to describe how they are feeling. Red for anger, Yellow for frustration or silly behaviour, Green for calm learning, and Blue for sad or tired.

For my children I used a similar idea. I liked to get them to think that they are in a different room for each emotion.

- Red Room - represents anger or rage
- Blue Room - sad or tired
- Yellow Room - frustration or silly behaviour
- Green Room - calm, fun energy

WHICH ROOM ARE YOU IN?

This made it easy for my children to exit one room and enter a new room which represents a change of their emotions. They found the visualisation easier. Going from one room to another. I could ask them "Are you in the blue room?" when they were tired or upset. Then I would ask them to go into the green room which we associated with happy and rested feelings. It was easy for them to think of walking from a blue room into a green room. I would often say "Can you come out of your red room, or do you need some space?". This was great as it gave them the language to use and gain control of how they were feeling. It helped them to regulate their emotions and work out how they were going to change their mood.

If you are not sure about these ideas, there are many others that are similar and might fit your family or children better. I have seen a system of icon's from happy to sad which can be used. Another idea is a colour wheel. It really doesn't matter which system you choose, you just need to be consistent and stick to it.

Consistency really is your friend here.

Intermediate Skill Stage (5 - 10 years):

Children of this age start to understand others points of view and can have reasoned arguments. They learn impulse control and have a deeper sense of empathy and compassion for others. These developments mean that they can establish and maintain positive friendships, although these tend to fluctuate depending on their interests.

The most important thing to think about for this age group is that ***no matter what system you did early on, keep it going.*** Don't change the system, just change the delivery as the children get older. You can now use more complex language describing the situation and the cause. When my daughter was upset about a friendship, I would take her for a walk and we would talk about what makes good friendships and if the other person is being a good friend. We also talked about when to walk away from friends if they are not making good choices. The hardest thing to do as a parent is after the conversations, let them make the decision on what to do.Watching them make the wrong decision is heartbreaking and you need to be there to pick up the pieces. Allowing them to make these mistakes early on means that they are more inclined to remember the lesson and make better choices when they are older. I had the best feeling when one of my teenagers rung me from a party to say to come and pick them up because their friends are not making good choices. I didn't question what those choices were, but turned up and took her home. I was so proud. This is when I knew all those long chats had worked.

During this age children are asked to write stories at school. It is a great idea to describe emotions in their proper names at this age so that they can use this language in their stories. They will use them and describe them in detail which makes their writing so much more interesting. It's amazing how they remember all those conversations you have had with them. When they are in times of extreme stress and during the teenage years, you can help them to work their way through the emotions especially if they have the language to describe it to you.

Supporting Emotional Intelligence through movies

I loved supplementing my conversations with movies as they can be powerful teaching tools. Disney's *Inside Out* is one of my favourites. It follows a girl navigating a move to a new city, with characters representing her emotions—joy, sadness, anger, fear, and disgust, sitting in the control room of her head. As the children watch they start to understand the different emotions and what they make you feel. They also see the battle which takes place between them in her head. Watching together and then talking about the characters gives children a safe, non-confrontational way to discuss their own feelings. Later, when they're upset, you can ask, *"Is Sadness taking over right now?"* It's a gentle way to start important conversations in a non-confrontational way. Watching the movie and talking about it helps children not to feel so alone when they experience a similar emotion to the character in the movie. These discussions help children to be able to articulate how they feel to you which alone will change their mental health for the better, now and into the future.

> 💡Parenting Tip: It is important to note here that you should not have a big discussion about how they could have fixed the problem or how they are feeling in the middle of the episode. Wait until everything has calmed down and you are both in a better frame of mind to have these conversations. It might seem like you need to let them straight away, but if they are in the height of an emotion they won't be able to fully listen or learn. A psychologist once told me that children only listen to the first 3 words you say when they are emotional. So if you start with "Johnny I don't …." that is all they are going to hear. Start with short phrases like "No. Not appropriate", then they get the message straight away. Save the 'why' later when they are calm and in their green room.

Advanced Skill Stage (10 + years):

Once they are teenagers we are hoping that they have these emotional intelligence skills mastered. Unfortunately teenagers have a great reputation of rebellion and causing frustration in adults. Although they may have mastered their emotions when they were younger, it is a new playing field now that they are teenagers. They need to learn it all over again through a process which rewires and strengthens our brain called **neuroplasticity,** the ability of the brain to create new pathways.

Lets get a little technical here while I explain the concept of neuroplasticity. No need to memorise all the scientific names but I hope you get a good understanding on how the brain learns and 'saves' its information. Every time you learn a skill you create a new synapse (pathway) in your brain. When you first complete a task it is like you have walked once through a field in your brain.

When you look back after one trek, you can hardly see the path you took. If you walk that same path 100 times, then you will look back and see that it has now become a dirt trek.

The more you do the task and use your skills the more you have walked the path. After walking 1000 times, it becomes a dirt road and now you can go much much faster on your road compared to the first time stepping through the long grass.

The path you create across the field is reinforced in the brain when it creates a substance called myelin forming a protection around the synapse (pathway) allowing easier and faster access for your brain. When you have done that skill 10,000 times, then it becomes like a 3 lane highway which you can do 250km/hr on. So, the more you use that skill, the easier it is to accomplish and re-learn when you are a teenager. That's where the saying "Practice makes perfect comes from", However I like the saying

"Practice becomes permanent"

Luckily, once you have learnt how to control your emotions, the second time you 'walk the pathway' it is easier. Specifically in the case of teenagers, we come across one additional hurdle. Hormones. Even the best teenagers will push the boundaries and rules at some time. You may ask why this is, we just learned that repetition is key? Well Teenagers get an additional boost of hormones as they grow and navigate through this time as a study submitted to the National Library of Medicine on the Maturation of the adolescent brain (2013) explored. These hormones help the brain to re-wire but make it very enticing to go against the rules and gain that adrenaline boost. Teenagers are getting ready for adulthood and need to explore the basic social skills they will need as an adult, thus the rebellion. You'll be happy to know that exploring these skills takes much less time the more repetition they have had in the past.

I had the best teenagers, however they did have their challenges. In a psychologists lecture I once went to, he mentioned that children of 12 have similar challenges in their emotions as 2 year olds. So if your little one has had lots of tantrums at 2 and they didn't learn how to get their emotion's under control, then it is 10 times worse at 12. Subsequently 14 year olds have the same learnings of a 4 year old and so on. I didn't think that this could be true, but as my children passed through the teenage years, I saw similarities in these ages. The 3 year old terrors I experienced definitely reared it's ugly head at 13. Thankfully, because we had worked on emotional skills early, those rocky phases passed more quickly. Believe me it is best to deal with the emotions at 2,3,4 and 5 years instead of 12,13,14 and 15 when they are big, more independent and influenced by peers rather than you.

Research on how the brain changes from child to young adult has found that up to 21 years of age the brain will remodel significantly (Robinson 2009). The unused connections or pathways in the thinking and processing part of the brain are lost, ready for more growth in other areas (Supekar, Musen, Manon 2009). This growth continues to occur from the back of the brain to the front during this time and children can seem volatile and have lost the lessons that they once knew so well. During these years they may have to be continually reminded of life and social lessons as the brain remodels and re-wires, but the skills come back easier each time. The brain won't just cut a pathway off forever, this would take a very long time and be of no use. If you started with your highway after only a few years and some struggles it may have turned back into a dirt road but its still much more drivable than that grassy field.

All the work is worth it and there is a big sigh of relief when you see them come out of those teenage years. The pay offs are worth the struggles when you see how your children are making good friend choices and good life choices in their 20's and beyond.

Now for some more detail about how you can help your child through the stages of learning about Emotional Intelligence.

Six Aspects of Emotional Intelligence

To help children thrive I would like to focus on 6 main aspects of emotional intelligence. It enables them to express themselves clearly, work through issues, as well as understand others points of view. It helps them to make good choices for friendships and have good relationships with a variety of different people including teachers, peers, elders and other adults. This can be an amazing advantage in the playground and later in the boardroom. These are:

- Self-awareness: Recognising your own emotions and how they affect your thoughts and behaviour.
- Self-worth: Understanding that you are worthy of love, respect and good things happening to you. Your inner voice is formed through what your parents say to you repeatedly. Make sure you say positive acclamations to your children regularly. Things like - "You are amazing", "Hello gorgeous girl", "Very clever of you", "Love that attitude".
- Self-regulation: Managing your emotions and impulses to adapt to different situations.
- Social awareness: Understanding the emotions of others and empathising with them.
- Grit: The ability to persist through challenges and setbacks while controlling your emotions when things don't work out and keep on trying until they do.
- Relationship management: Using your understanding of emotions to build and maintain healthy relationships.

Teaching these from an early age helps children solve problems, navigate disagreements, and often step into leadership roles. Emotional intelligence supports better mental health, reduces stress, and prepares them to thrive in adulthood.

The Power of Self-Worth

It is worth talking a little more about **self worth**, which by definition is how you think about yourself, how you value yourself, how much you value your own ideas and how deserving you feel of love and respect. Whereas, **self-esteem** is about external measures of your confidence in your abilities and achievements. People who have high self worth have their **inner voice** telling them that they deserve good things happening to them in their life. This inner voice is created by what our parents say to us repeatedly as children. I bet you have heard words in your head which you know have been said by your parents. Sometimes you don't realise that it is their voice you hear. Those repeated words like "gorgeous", "clever", "beautiful", "clumsy" are what we end up saying to ourselves and believing to be true. They have become our inner voice.

"Our parents words become our inner voice"

The Inner Voice is a powerful voice and can influence our adult lives in positive and negative ways. One of my most moving parenting moments can when I asked my daughter what she thought when I called her "beautiful girl". I had intentionally called my daughters "beautiful girl" or "gorgeous girl" when they come down in the morning and throughout the day. I started to worry that I might be unintentionally giving them a body image problem (a bit late as I had been doing this for over 12 years). Her answer was a glimpse into what wonderful adults they had become. She said "Oh, that means that I have a beautiful soul". I was about to ask why when she very nonchalantly said "That's because you value what I'm like as a person more than my appearance, so I know that you mean that I have a beautiful soul when you say beautiful girl". It was a tear-jerking moment for me. I bit my lip and sat in the car after she got out for a good 5 minutes and cried happy tears. I had seen first hand that this simple act of telling my girls everyday that they were gorgeous and beautiful helped their positive self-worth, self-esteem and helped to build their emotional intelligence into confident, kind young adults who are loved and are worthy of good things happening to them.

Practical Tools for Managing Emotions

Trying to teach children how to manage emotions seems very hard. Especially when you are in the middle of a tantrum. It is good to have a few ideas in your back pocket to bring out in the heat of the moment. These are great strategies that help children work through difficult feelings:

- Hold your breath: Get them to breath in for 4 seconds, hold for 4 seconds, and then out for 4 seconds.
- Safe Space: Have a designated safe spot they can go to calm down until they are ready to talk. With my children we use to have designated spots like in a car, another room or any quiet space, where I would sit with them silently until they were ready to talk to me.
- Trace a 8 figure on your wrist: Have them trace an 8 symbol on their wrist or hand to focus and self-sooth. It gives them focus on a task and they can feel something tactile so they get distracted and calm down.
- Cry: It's important that they know it is ok to cry. It's a release of emotions which is completely normal. As a parent our hearts break when we see our children cry. Sometimes being present is all it takes. I remember when my teenager daughter was really upset about some friend issues. I sat outside her door for hours listening to her cry. I would knock on her door every now and then so she knew I was there. Finally she stopped, walked outside the door, hugged me and said "It's going to be ok Mum".

That was all I heard about it. I didn't need to know the details, she knew I was there to support her and that was all that was needed.

- Consistency and Stability: Remember that you are their mentor, there coach and their stability. You have the answers to their problems. By being consistent with your rules you make it easier for them to follow your example and suggestions. An example of clear consistent firm rules are no biting, no kicking and no hitting, no matter what. These rules build safety, and trust so that your little one can learn best.

I have a funny story about biting. It's one I will never forget as I was trying my best as a parent and feel that I failed miserably. I was reading a book, Toddler Taming and remember it saying that if your child bites you, you do the following steps; 1st bite: say no; 2nd bite: slap their hand and put them in a corner for time out; 3rd bite: put them in their room and close the door. Well, I had this information in my head and the first time my daughter bit me on the bum I got such a shock that I said "NO", slapped her hand and put her in her room and closed the door. Then it occurred to me that it was a step by step process. Whoops! She got such a shock that it actually worked and she never bit me again.

A strong consistent approach with no room for excuses will work every time, but it must might be 100% of the time. Make your life easier by being firm and consistent.

Naming Emotions

It can be hard for children to be able to understand how they are feeling. When they are babies they cry to get attention to fulfil their needs. When they are toddlers they learn words to help get across what they want. But as they grow older they need to be able to 'use their words' as I have heard many Mum's tell their little ones. What the emotion is can be hard for them to articulate. Here are some ideas on how you can help your child let you know how they are feeling and also how to work through those feelings;

Children need to understand the name of what they are feeling.

Saying "I can see your feeling angry at the moment" can help them label their emotions. As we discussed earlier in this chapter, I like to put a room colour to the emotion - "You look like you are in your blue room, are you feeling sad?" The emotions I put to each room are as follows;

- Red room - angry, frustrated, very heightened, and fast movements.
- Yellow room - excited, laughing out of control, silly and bored.
- Green room - calm, happy and controlled.
- Blue room - sad, upset, worried, tired.

They need to be able to name what it feels like.

"How are you feeling in your body?", "What feels different?","What colour room are you in at the moment?". Some examples of how your body feels

- Angry - hot head, clenched fists, wanting to hit or scream, stomping, scowl,
- Frustrated - sick tummy, scrunched up face, punching
- Excited - buzzy feeling, jumping, moving fast, smiling, laughing
- Sad - crying, pout, shoulders slumped, walking slowly, head slumped

They also need to be able to move from emotion to emotion.

"Can you go from your red room to your green room" They might need some time or techniques (See how to help your little one with tantrums above for hints) to move from red or blue room to the green room and that is ok. The more they understand this, the easier it gets and the more emotional intelligence they develop.

The idea is to put these skills in place just as life happens. It's not a 'lesson' as such, just mentioning the ideas and teaching the skills as life goes on. Ensuring that they understand the concept will require a lot of talking and maybe some drawing and colouring, but after that it is just on the fly. That is essentially you mentoring your child. Having these skills in place when they are calm helps when everything goes off the rails as it is embedded in their routines.

Praise and Correction

So many times I saw other children in the shops walking alongside their parents and being well behaved when I would be continuously saying "that's not appropriate". I would wonder how they got their children to behave so well. How did they encourage them to act in a way which works within the community and our lives? Finally I found what worked best for me and lots of other parents I have helped. It is praise and consistent correction.

Praise is a good way of encouraging those positive behaviours. Guiding your children through praise can be a positive way to re-enforce the behaviours we want them to have. When they are walking quietly beside you in the shops being little angels a quick *"I love how calm you are now"* is all that is needed. When you see them act in an empathetic way to others, just a simple genuine comment works best. *"Oh I loved how you gave Kate a hug when she was sad"*. That is all that is needed. Not a long convoluted conversation. Not a huge explosion of fireworks, just a quick comment which is easy for them to digest.

Praise can also be a double edged sword. Too much can be damaging to their self-esteem and self-worth but not enough is just as bad. children are clever and know when praise is earned and when it is worthless. How do we know if we are praising enough or too little? Think of it like "the toilet test": If a child has pooped in the toilet for the first time, then fireworks go off, lots of clapping and lots of praise is appropriate. However you are not going to praise a 4 year old for something they have long mastered. You don't even mention it, because if you did, it would be wrong.

This works for everything they do. Praise should be for the first few times they do something, when they have worked hard towards a goal, or when they achieve something important. Don't praise for the sake of making them feel good, because it becomes mute and damages the image they have of themselves. By all means support them and say you are proud of how hard they worked or how they kept trying, but don't praise the outcome if there was no effort put in. The funny thing about praise is that some children need only a little praise to understand that they have done a great job or what is acceptable behaviour and others need more repetition and guidance.

Boundaries and Consequences

Currency

Children need boundaries to feel secure. Some parents resist this, but freedom without limits often leads to chaos. Think of traffic laws: we stop at red lights for safety, not restriction. In the same way, clear family rules guide children to thrive within safe boundaries. I have had parents say that they want their children to be innovative and not be restricted by boundaries, however I point out that you need to know how to work within the boundaries before you can successfully work outside them.

When behaviour slips or we want our children to do something for us we can use their 'currency'. It is something meaningful to your child as motivation. It could be a toy, favourite show or as simple as a visit to the park or a picnic together. It is also important that if they refuse to do the behaviour or action there has to be a consequence which is followed through 100% of the time.

I remember when I tried to find a currency for my children. I tried everything. My friend had their child's favourite show, another friend had sultanas. Her child would do anything for a sultana. But I had nothing.I followed the consequences as I was told to do and at one stage one of my daughters ended up with an empty bedroom only containing a bed, pillow and blankets. I just wanted her to pick up her toys. It was very simple but she ended up with all her toys, her books and even her favourite dress put away out of reach. I just couldn't find her currency. That's when I found out about how you can **make** a currency for children. And it was all about consistency and using the same currency, whether they are actually motivated by it or not, they understand what it means. That is the key and it changed my life. I had a toy which I gave to her each time she earned it and I had a timeout area which was a boring space for her to go for her consequence. The FOMO was real and she did anything to avoid the timeout area. You ask why it works? It is because children's growing brains need structure so that they can learn and grow. This routine gave her the boundaries. As long as your boundaries are 100% consistent, it doesn't matter where those boundaries are, you will have less behaviour issues.

By giving consistent boundaries to children early, they feel safe and can start to learn to work around those boundaries, within the rules, to achieve their own outcomes. This is an important lesson to learn for life. It's as simple as, children need to learn how to behave in society. The fact of the matter is that we live in a world full of boundaries. You cannot run a red light without a consequence. We understand why the law is there and we abide by it for everyone to be safe. Even animals such as chimps, dolphins, dogs, wolves, gorillas, zebra and elephants, are guided by their parents about how to live in their communities. Their mothers reprimand them when they go against the rules of the community. No young are allowed to do whatever they like, it's unnatural and usually unsafe. They are either guided by the community or life gives them the rules through hard lessons. Guidance means that they can miss the hard life lessons and learn in a softer, happier way.

Incentives

I always said that incentives are rife in my household. I would give my children incentives to do things that they didn't necessarily want to do. While some might call it bribery, I see it as teaching them to work toward goals. To teach this lesson to my own children I chose to get them to learn an instrument (as I enjoyed so much as a kid myself). You could also choose a sport or dance. This was a big ask for a child and requires consistently practising everyday. The long term benefits such as; learning how to learn, resilience, creativity, and supporting their brain development were so worth it in my eyes as I knew it was going to be worth the struggle. On a side note, music is the only skill which uses all the hemispheres of the brain at one time. Anyway, as far as my 4 year old was concerned, all I was doing was asking them to do tricky things everyday. They couldn't see that far into the future or appreciate all the benefits they would get as they got older at school. Actually they didn't care at all about the future. They just wanted to play. So the solution was that I would give them their own reasons to practise, ie. provide them with an incentive.

To create an incentive I would get them to earn a toy, or an experience like a picnic. 3 practises equalled a picnic, 5 practises was a new toy. Sometimes I would break up toy sets or lego sets and give a piece of it everyday if they were finding practises particularly difficult. They could see 3 days into the future so were happy to work towards this goal but sometimes it needed to be daily. Without the incentive they would have to understand how they were going to feel 5 years into the future or even further into their adult lives. This is impossible for a child to do.

I remember when my daughter came home to me in year 2 and said "Hey Mum, if I read everyday, like I practise my flute, I will get better and better and it will get easier. That's right isn't it?" It was at this point that I realised that my incentives were working. She knew that her everyday practise benefited her and made life easier. In a few months she started reading

Harry Potter which her teachers were astounded at. She had learned how to learn. I felt all those days of sitting and listening to flute music was worth it.

Once she became good at the instrument it became it's own reward. She started to see success, and motivation was much easier, although my bank account would disagree at the times.

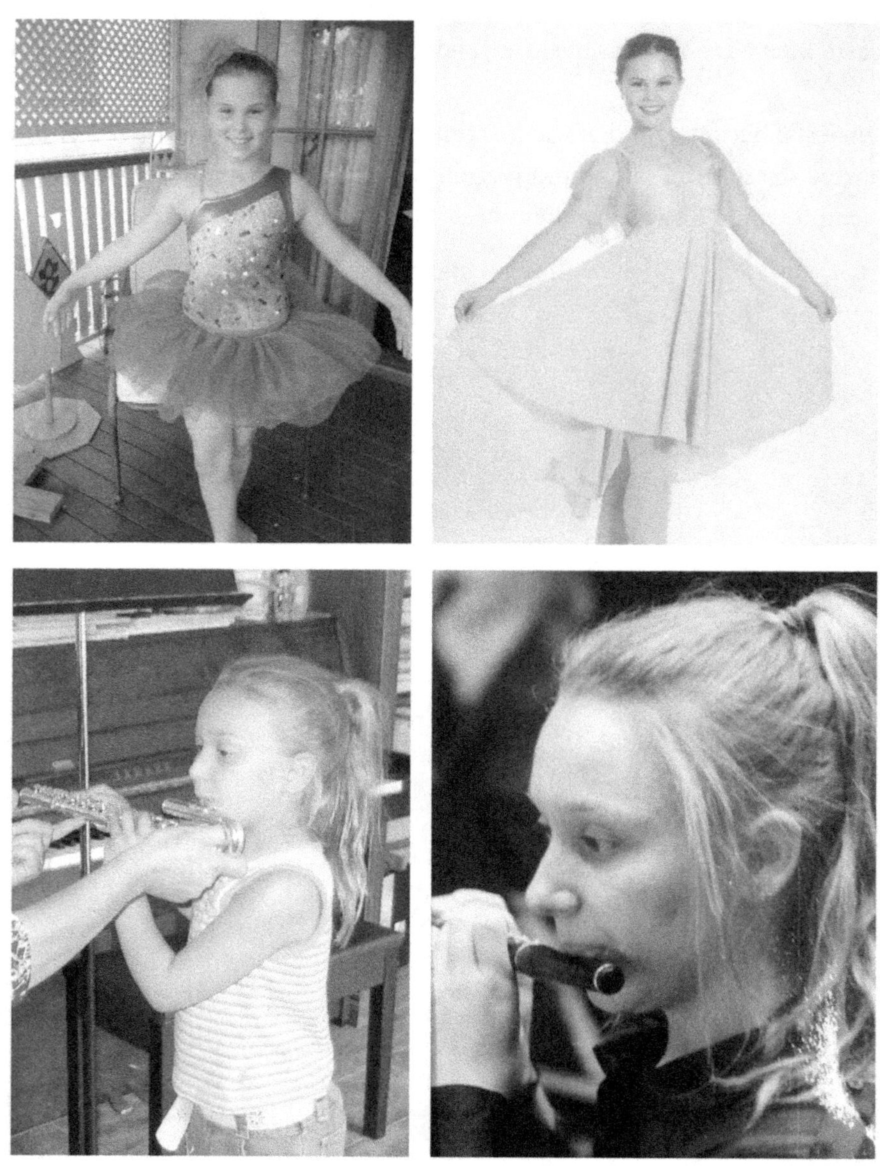

Parenting Tip: If you are asking your children to do something for you, which is good for their future, give them an incentive now.

No-one works for free, we get paid. Some people work for money, others like volunteers do their jobs to make a better life for others, which in turn makes them feel good.That's what I mean about being paid with feelings. children deserve the same.

Emotional Intelligence Summary

Developing Emotional Intelligence takes years of practice and patience. It requires consistent correction over many years. As children's brains change and grow, they learn and need to re-learn social and emotional skills. Your job as a parent it to provide the framework and guidance through consistency, boundaries, conversations, repetition and experience, so that they can successfully enter adulthood with good values, strong emotional intelligence, and the social skills they need for their adult lives.

If there is anything I have learnt from how children learn best, it is that consistency, boundaries and guided conversations make for a calm, social child who can handle stress and anxiety like a pro. They learn by watching how we handle challenges, by talking with us about their feelings, and most importantly, by practising over and over. With your guidance, they'll enter adulthood equipped with the values, resilience, and emotional skills they need to flourish in a world where AI can do many things—but never replace human connection.

SOCIAL AND LIVING SKILLS

Firstly, what exactly are social and living skills and why do children need to know about it? Some people say that children are little free beings who shouldn't be burdened with the knowledge of social and living skills.*"Keep them unburdened by the complexities of society and life. They will learn about life too soon anyway"*. Alternatively, I say that if they learn about how society works early they are better equipped to cope with all that life brings. Further, they will develop empathy and kindness early on, a sure way to becoming a successful adult. Social skills are how we talk and interact with others. To make the most out of life and relationships it's important that children understand how to treat others as well as how their actions and words affect those around them.

There are 3 main areas of social skills:

- Verbal and Nonverbal communication
- Co-operation and conflict resolution
- Empathy and understanding social cues

Teaching Social Cues

Depending on the age of the child we can teach them about some of the social cues which are easy to identify. Body language such as frowning, or smiling, or even just laughing. When they are little we can talk about how they and others, show their emotions. If you are happy, you have a smile on your face, laughing and have a high voice. If you are sad, your shoulder sag, you frown and sometimes you can crying.

💡Parenting Tip: If you are waiting for something or on public transport you can quietly ask your child if they can see someone who is happy in the bus. They tell you who it is by describing their clothes and not pointing.This teaches discretion and body language at the same time.As they get older you can develop this by getting them to remember who at the restaurant was angry or having a fight with another person at a table. You can have these discussions in the car or on your walk home. This also gives them something to do while waiting for food, other than watching screens or playing on a device.The more detail they can remember the better. You are sharpening their observation skills, building empathy and sparking conversations without the use of screens to fill in the time. Just watching people is also a great way to build knowledge about body language.

Pretend Play and Mimicking

Pretend play is a powerful way that children process the world around them. It is also another great avenue for adults to help little children process situations, understand social cues and learn conflict resolution. Have you ever listened in on your children playing. It's so fascinating listening to them process different situations. They often try different avenues of conversation to see what the result is.It's trial and error without the pressure of real life, experimenting with social outcomes. This is why imaginative play is so important as it helps children to process the world around them in a safe space.

Mimicking is equally an important tool to learn about the world around them. They get to feel what it is like in their bodies and practise people skills. I bet you have seen your children wanting to mimic you cleaning with their pretend broom, or toy broom. Its both cute and is also so important for their development. We are after all developing a growing mind which can take 25 years to fully mature (Arain et al, 2013).

Children need these no pressure situations, to process how our complex world works and to practice these people skills. Hands on play is the best way to learn. If you join in on the play you can sometimes guide them in what is appropriate people skills and what is not acceptable. When my daughter said to me "That's not appropriate Mum", one of my friends asked how she knew such a big word and I realised how much she was watching and taking notice of me. My words were becoming hers. It was clear that I had said this way too many times, but she used it correctly and really knew what it meant, so she was learning. Those little comments and talks go a long way to helping children develop their own social skills. The most important advice to remember when talking with your children is that;

"Your words become their inner voice; so make it positive"

I found this to be so true when I heard my dad's sayings popping into my head. When it first happened to me I couldn't believe it, but then I realised the wisdom in those words. Those positive affirmations I gave my children everyday work. They believed the words I was saying and they lived by those values. This was the language going through their minds. It was such a privilege to hear my legacy, my words guiding them throughout their lives even when I wasn't there. My most precious gift to them.

Lessons from Dale Carnegie

There is a very old book *"How to win friends and influence people"* by Dale Carnegie (1936). Though written nearly a century ago it's sales are over 30 million and it is still selling strong today with it's principles remaining just as relevant today as it was in the early 20th century. People continue to use his techniques in the boardroom, and we can teach our children to use them in the classroom and playground.

Dale talks about how to get the most out of situations and help people, who you work with, to reach their own potential. He discusses how to build colleagues up so that you can both succeed. This **collaborative approach** is essential in the emerging world of AI. Working together to the same end can boost up everyone. Actually I think these skills are more important in the classroom than the boardroom, coming from a teacher.

Here are some of my favourite things discussed in the book to teach your young person:

Be their role model:

When dealing with people the best approach is to not criticise, condemn or complain, no matter how much you want to. This will ensure that people are feeling good about the relationship and they will do their best to make it work.

How does this look for a child is simple. Be kind.If you can't say something kind then say nothing at all. If you hear your child say hurtful things, ask them if it was 'kind words' they were saying. If it wasn't, then that is a great teaching opportunity to talk about feelings and how your words can effect others. Make sure they know the difference between kind and helpful words vs mean and teasing words. It should only take a minute to have this discussion. We are not lecturing our children, just guiding them. This quote from the book "How to win friends and influence people" still stands true today.

> *"Criticism is futile because it puts a person on the defensive and usually makes him strive to justify himself. Criticism is dangerous, because it wounds a person's precious pride, hurts his sense of importance, and arouses resentment. .... Any fool can criticise, condemn and complain—and most fools do. But it takes character and self-control to be understanding and forgiving."*
>   *- How to win friends and influence people 1936*

Appreciation goes a long way;

Give honest and sincere appreciation for what people do. Say thank you when people help you or aide you. Showing your children that gratitude is important, but more importantly being sincere, is essential. They will mimic you, so make sure you put your best foot forward and be grateful for even the smallest things. Reduce the complaining to a minimum. It can be harder us all to do complain less, but it really changes your mindset and reduces stress. It is an important skill to help your child become resilient and happy.

> 💡Parenting Tip: Talking to friends about what you are grateful for, when your children are in ear shot is a great way for you to reinforce that this is important to you. Remember we have talked about how important repetition is and how children innately want to please us. This is fabulous opportunity to use both these super powers we have as parents to help our children live kind, successful lives.

Positive influence

Be that positive influence with your friends. Telling people you appreciate what they do, or that you are grateful for their input. It is important that you are sincere in all these interactions to build trust in your relationships. The more you are a positive influence, the more people will take you serious.

Sometimes it can be hard to find something positive and often it will fall on deaf ears, but making sure that you are the positive influence with the people around you makes your heartfelt words of appreciation received in the best possible light. It's all about reputation, gratitude and kindness. They say action speaks louder than words and they are correct. Having a reputation for complaining about everything means that when you start being positive or appreciate someone, they will always look for the alternative motive and ask themselves, "What does this person want from me". It's important that teenagers understand this concept and it can be taught through discussions and mentoring from the age of 10.

Honest and caring

People who are kind, caring and honest have better relationships and more success in life. This statement has been verified by many studies over the years. I have already mentioned a few. They have found that most people want to be friends with the kind people and are willing to help them to succeed. There ability to make everyone feel good about themselves and the tasks gave them a superpower. They would make people want the same thing as they wanted themselves just through being an honest, kind, caring person. It all came down to the way they made people feel.AI cannot compete with humans in this area. Teaching the importance of these traits to our children from a young age by example is the best way for them to see the positive effect this can have on the people around you.

Friendship Tips for children

I hear from so many parents who want to know how to help their children with friendships. children go through so many stages of development at different times and at different ages friendships can be hard. Some children understand the nuance of social cues and others are still learning. Friendships will change throughout the years and one second your child will be best friends and then the next day they can be mortal enemies with the same person. That is normal with children. You still need to help them navigate through this turmoil so that they can make good friends as young adults.

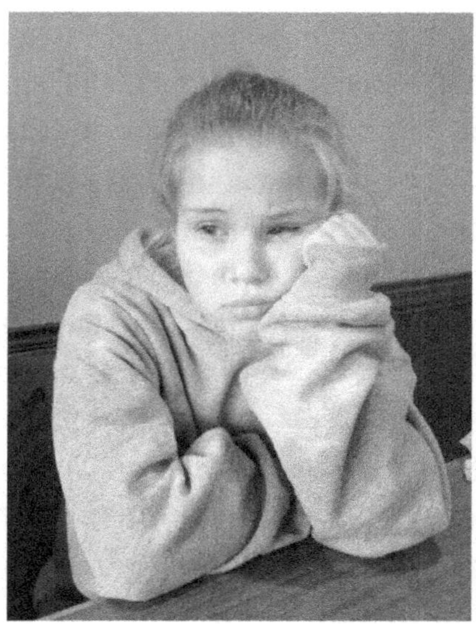

When I hear children say *"I don't have any friends"* my heart sinks. I would love to just wrap them all up in cotton wool and protect them from the hurt, but let's face it, children can be cruel. Knowing that relationships can change so quickly during childhood here are some practical tips on helping your children make new friends. (Tip: It works with adults too)

- Become genuinely interested in other people. Help your child to ask them about the other person and teach them how to continue asking questions about what the other person loves. Start with questions such as "What did you do on the weekend?". Help your children to find what other people are interested in and have a passion for. Teach them to see the body queues such as an excited look on the other persons face, or the voice going up when they are excited about a subject. Then they can start a conversation about that topic.
- Smile a lot. Smiles and giggles are contagious. I saw recently on Instagram and TikTok a guy laughing at his phone on the train. At the end of 2 minutes, most people were smiling and laughing for no

reason. It made people happy, just listening to the laughing.

💡Parenting Tip: Have your children smile at people on a walk and remember how many people started smiling back. It's a great way to show how you can influence how people feel in a positive way.

- Remember a person's name. People love hearing their names. Use it and they will remember you. It works because it shows that you care about them.

💡Parenting Tip: Have children remember all the names of people at a party or event you go to.During the event go up to the children and see how many names they can remember. The more you do this, the more they will anticipate it and make sure they can please you by remembering the most. My children were very competitive so I would make this a game between them.

- Being a good listener will encourage others to talk about themselves. People love to talk about their lives. Knowing what questions to ask to be interested in others will help children to keep friendships.
- Talk in terms of the other person's interests. Find a common interest and start talking about it. People love talking about their own self so take advantage of this and find common ground.

💡Parenting Tip: Most children have a common ground in the school or sporting team they are in. Talk to them about finding what they both like eg. they may have the same love of the oval and soccer, or the playground and the monkey bars, or even the library and board games. These common grounds are great to start the conversations and build friendships. Some of my children's friends still get together regularly to play board games together or have powerpoint nights.

- Make the other person feel important.Firstly, to do this sincerely you have to believe it yourself. If you don't, then don't try it as it will backfire on you. To tell you the truth, most people find this really hard, but it is worth looking into yourself and making the change to really make others feel important in your life. It can also open up doors which would normally be closed and give you opportunities which you might never have seen.

🔅Parenting Tip: To help your children to do make others feel important have them remember a little detail about them. This will the other person the sincerity of their friendship and start the journey of friendship.

- Understand who is a good friend and who you should walk away from is the most important skill you can teach your child. Believe me, it will take many years to perfect it and many tears. It's about understanding that actions speak louder than words and you cannot control what other people do and say. It's helping them to understand that they have the power to leave friendships which are toxic or don't advantage them.

💡Parenting Tip: When the friendship is toxic, give them the advice and then let them make the decisions. They need to make the mistakes to be able to make better choices later on. The hardest thing as a parent is to watch your child make a bad friendship where they are taken advantage of. Guide them to understand that it is wrong. You may have to watch as they decide to keep the friendship, only to see the aftermath of tears or worse. All you can do is be there to support them through the heartache, but that is enough. They need to learn for themselves.

Persuasive Skills

Part of understanding people skills is learning how to persuade people to our way of thinking. Persuasive writing is taught in school and for some children it comes naturally but to others they have to really work at it. The world is made up of lots of different people and let's face it, it would be boring if we were all the same.It's important for children to know how to persuade someone to help out or to come around to their way of thinking. These are natural skills used by communities to work and live together.

It's important for children to recognise the difference between persuasion and manipulation. If you have persuaded or convinced someone honestly that your way is the best, then that's persuasion and you can let the change happen. If they don't want to make the change and you walk away, this is also healthy. So, persuasion uses honest reasoning while respecting the other persons autonomy. It becomes manipulation if you use your persuasive techniques to change peoples behaviour for your positive gain and their detriment. It is often when people use deception and underhanded or exploitative tactics where this becomes damaging and is a negative manipulative tactic which should be avoided at all costs.

Persuasion is a powerful skill, however knowing when to convince and when to walk away is a superpower. Understanding that not everyone can make the changes is important. It's ok to move on and walk away. This is a very important lesson for children. Remember it's not about who is best, but how we can all help everyone achieve a common goal. It is called a 'win win' situation. Children who can do this have the makings of a true leader. Teaching persuasion early on in your children lives can be a real game changer for their future.

Respecting Others

Respect is a word which is bandied around a lot at schools. What does it actually mean? How can children show respect for others? I often ask children "What does respect look like?" And they come back with answers such as, listening to others, letting others talk, keeping our hands and feet to ourselves, being kind. It is very interesting as these are staple rules which we are asking school children to follow at school. So they are interpreting respect as following the social rules. But it is more than that. It is regarding the feelings, wishes and rights of others. It can also be a feeling of admiration for someone else because of their abilities or achievements. So it is all those things the children said and more. Helping your child to know what respect "looks like" in terms of behaviour and "feels like" when others respect them, can protect them against many injustices. Alternatively, helping them to know what disrespect is and that it is ok to walk away from people who continuously disrespect you is just as important, even more important I think.

When we judge people and say that they have a good character, it is often about how they treat others and respect others. Thomas Carlyle, a famous psychologist says: *"A great man shows his greatness by the way he treats little men."* He means that the way we treat others who help us, shows our good character. Whether it's a work colleague, a cleaner, a barista or a waiter, showing kindness and respect costs us nothing but will go along way to building your good character. This kindness can be as simple as "Hi, how are you?", instead of ignoring people who work for you. It doesn't take any time but it does make people feel valued and seen. People will then go the extra mile for you and do their best to help you out when they can.

Showing children how to be truely grateful for the work people do builds the skills they will need to become a truely success leader in the future. I have lived by this philosophy since my early 20's and in every workplace, I have found that treating others with respect helps my work outcomes and makes others happier in the workplace. Showing gratitude, respect and kindness will help your children work with others in the AI world where the human connection is paramount to success.

Living Skills

Living Skills can include simple tasks such as learning to tie shoe laces, putting out the dishes for dinner or sweeping the floor. These tasks are important to get children involved in from the age of 2 years. Tasks will grow with your child. For example a 2 year old can easily put away their plate in the dishwasher when they are finished and put away their toys at the end of the night. Whereas a 12 year old can do the laundry, wash the dishes and even start to make a meal or two. I was 12 when I was making meals for my family and doing basic housework. This has held me in good stead all my life as I just quickly clean and make meals without a second thought, whereas some of my friends who were spared these chores as children still complain about it. It is important to note here that I don't agree with paying for children to do chores. My reasoning is that as a family you are working as a team. You all contribute to the mess, so you all contribute here you can for the working of the household.

Here is a table of what tasks I suggest for children to do at different ages:

AGE	TASK
2 - 3 years	Put dishes away at the sink or in the dishwasher. Pick up toys and books Put rubbish in the in Put their dirty clothes in the hamper Fold rags and tea towels Sweep with a toy broom (you might have to go over it)
4 - 5 years	Feed Pets with supervision Set and clear away the table Match socks and put away laundry in drawers Make bed Put away groceries depending on your kitchen height Empty the bin Sweep with a small broom Wipe over the bathroom with a cloth
6 - 11 years	Wash and fold laundry Sweep and mop the floors Vacuuming Take out the bin Wash the mirrors and wipe over the bathroom Clean rubbish out of the car Clean their room Clean joint use rooms such as the dining, kitchen, playroom Organise rooms such as playroom Simple gardening tasks such as weeding, watering
12+ years	Mow lawns Wash windows Iron Wash car Cook simple meals All laundry including washing, hanging and folding Sweep, vacuum and mop floors. Babysit

If you need any more reasons to get your children working here is a big one. The advantage of children experiencing these chores/tasks when they are little, means that they are also exposed to important concepts such as spacial awareness, chemical reactions and basic physics laws (like gravity and balance). In the new AI world, science, especially chemistry, physics and math, will be of the utmost importance.

Children can start to develop an understanding of our world through doing things like chores, playing and watching their parents (which use to be called mentoring). In the new emerging of AI, people will have an advantage if they can explain what is happening in scientific terms, using physics, chemistry and maths knowledge. Teaching children this doesn't have to be difficult. Using scientific words like gravity, chemical reactions, or even just talking about adding together, will help them to understand that you can explain what happens in the world with science.

If children don't experience life and be exposed to scientific terms, then the science doesn't mean as much when they learn it in detail at school. Hands-on is one of the best ways to teach children and it's fun to boot.

Fun Activities to support Social and Living Skills

◑◑ **People-watching games:** Spot emotions in public and talk about body language.

- **Skills Built:** Observation, emotional intelligence, empathy, and non-verbal communication.
- *AI-World Connection:* In an era where machines can read data but not emotions, the ability to notice subtle cues e.g. like tone, posture, or micro-expressions. These give humans the ultimate edge. These observation skills translate into future roles in AI ethics, human–AI interface design, education, psychology, and leadership, where understanding people, not just systems, drives innovation.

🐾 **Role-play:** Act out playground scenarios to practise problem-solving.

- **Skills Built:** Conflict resolution, creative problem-solving, perspective-taking, and collaboration.

- *AI-World Connection:* When children practise resolving disagreements or testing solutions through play, they're developing the same adaptive reasoning that future negotiators, mediators, team leaders, and innovation consultants use when guiding human–AI teams. It nurtures flexibility, the hallmark of thriving in a world that changes as quickly as technology evolves.

♥ **Gratitude practice:** Share three things you're grateful for each day.

- **Skills Built:** Emotional regulation, optimism, mindfulness, and empathy.
- *AI-World Connection:* Gratitude strengthens wellbeing and resilience which are qualities AI cannot replicate but every successful innovator needs. In a future full of automation and high-pressure digital environments, emotionally balanced humans will lead teams and create healthier workplaces. These traits underpin careers in wellness leadership, education, healthcare, and ethical AI development.

Friendship Role-play: Practise introducing yourself, remembering names, and asking questions.

- **Skills Built:** Communication, confidence, active listening, and memory skills.
- *AI-World Connection:* As children learn to introduce themselves and build connections, they're laying the groundwork for networking, leadership, and collaboration in the digital age. These interpersonal skills will empower future entrepreneurs, project managers, community organisers, and cross-cultural communicators working alongside intelligent systems that still rely on human connection.

⚖️ **Mini debates:** Take turns defending silly topics (e.g., "Which is better—ice cream or cake?").

- **Skills Built:** Critical thinking, persuasive communication, reasoning, and respectful discourse.
- *AI-World Connection:* Debating trains children to analyse information, build logical arguments, and defend positions with evidence, all essential in a world of misinformation and AI-generated content. These skills connect directly to future roles in law, journalism, data ethics, AI communication, and public policy, where clear human reasoning must guide the machines.

AGE	ACTIVITY	EXAMPLE
1 – 5 years	Pretend Play	Build a Fort, dolls house, cubby
	Dress ups	Pirate clothes, animal tails and ears, princess dresses and crowns, ribbons, coats, hats, teddies, dolls
	Toys	Pretend kitchen, toy tools, toy cleaning equipment, dolls, teddies
5 – 10 years	Board Games	Guess Who, Cadoo, Jenga, Uno, Chameleon, Herd Mentality, Social Bingo, Cranium
	Role Playing	Having dress up's like nurses uniform, doctor bag, hard hat, policeman hat, etc.
	Games	Charades, Pictionary, card games, Two truths and a lie game
	Lego building	Build a creation together, either from instructions or collaboratively make your own.
	Puzzles	300 + pieces
10 + years	Board Games	As above + Codenames, The Chameleon, Deception
	Puzzle race	See who can put the most pieces down in 30 minutes.

♟ DEBATING SKILLS: TEACHING CHILDREN TO THINK, REASON, AND INNOVATE

Debating is more than arguing a point. It's about learning to think critically, communicate clearly, and back up ideas with reasoning. These skills help children solve problems creatively, respect different perspectives, and innovate. All of which will be essential in a world shaped by AI. As you learned at the start of this book in my family, debating often began in the car during our annual road trips. To pass the hours, we'd choose topics and argue different sides. It was fun, but it also gave the children a chance to practise forming arguments and convincing me of their ideas. Not only was I being convinced to buy them a new toy but they began applying their debating skills to some of the problems they came across.

The one which still sticks with me is when they wanted to have a picnic on the flat roof of our veranda. Yes I said the roof! When they asked I was horrified. I thought I was being clever and said "You both can't possibly do that as you couldn't carry items up the ladder, its far too unsafe". I not only mistakenly didn't tell them outright no but also hadn't realised there was a loop hole in my comment. In 10 minutes they came to me, with big smiles on their faces. They had designed a pulley system to get the picnic things up on top of the roof. They would throw a rope on the roof and then attached the bucket to it. I couldn't say no. They had covered all bases, showed me they could manage the risks and weren't breaking any of my rules. So for the next few weeks, after school and with my supervision, they would have their afternoon tea on the roof of our veranda. I knew they were safe and they stayed within the rules I put in place. If they had been silly up there or it became unsafe they were told that it would stop. They had a ball every afternoon. There smiles were worth it.

My children had shown that they could think, reason and innovate within the boundaries set. All because I had started teaching them debating skills. They formulated all the reasons why and had answers to all the issues I raised. They had thought through the whole problem and could articulate their opinions and back them up. These are the skills which are needed to thrive in our changing AI world.

When to Start

Children can begin learning the basics of debating around age seven or eight. At this stage, they understand fairness and logic well enough to practise structure and reasoning. Remind them that a debate isn't an argument or a fight. It's a **structured discussion with rules**, and if you don't follow the rules, you lose the debate.

Using our annual road trips we would practice these skills. We would choose a topic and if they didn't agree with me, we went through how to form a structured argument to convince me their option is better.

Some of the rules I taught during the debates are as follows:

- Show respect for the other person's opinions. Never say, "You're wrong."
- If you are wrong, admit it quickly and emphatically.
- Begin in a friendly way.
- Get the other person saying "yes, yes" or nodding their head, then you know you have convinced them
- Let the other person do a great deal of the talking taking note of vital points you can rebutt.
- Letting the other person feel that the idea is his or hers can win the argument. Have you ever had an argument turned on you. It's very humbling but it is a skill which you can teach. It's all about critical thinking.
- Try honestly to see things from the other person's point of view so that you can analyse why yours is better. Make a list of pro's and con's. Maybe they do have the right choice, in which case, admit it and move on with no hard feelings. It is ok to be wrong, but only if you have explored why.
- Be sympathetic to the other person's ideas and desires but remember that you don't have to change your ideas just to fit in.
- Dramatise your ideas can help you to organise your thoughts and structure your argument.
- Throw down a challenge to the other person. This way they need to really think about your side as well and maybe they may realise on their own that you are right.

The next thing I taught them was how to lose a debate with grace:

- Admit that you are wrong and let it go
- Give your opponent a smile and make the change if one is needed
- Getting the right result is the more important than being right
- Emphasis the goal of the debate / question not the winning
- Don't resent the person who won

Process of debating

Start with Simple, Engaging Topics:

Choose topics that are relevant and interesting to children, like "Should we get a cats or a dog?", "Do we have Ice cream or cake for desert?", or "Should we go to Movie World or Sea World on the weekend". These topics allow for easy expression of opinions and build confidence in formulating arguments. Have a debate on what you are going to have for Sunday dinner or if you should do homework before or after dinner during the week. Whomever wins gets to have their choice. It is paramount that you follow through on these debate outcomes.

Introduce the Basic Debate Structure:

Explain the concept of a debate as a structured discussion with rules. Introduce the idea of having a "for" and "against" side, and the importance of supporting claims with reasons and evidence. Start with short opening statements and gradually introduce rebuttals and closing arguments. Emphasise the importance of taking turns and listening to the other side. Use visual aids or graphic organisers to help structure arguments.

The basic structure of the debate is as follows:

1st Affirmative	states their case and a few reasons why
1st Negative	discusses the rebuttal and then states their case and reasons
2nd Affirmative	discusses the next point and responds to the negative case and reasons
2nd Negative	discusses their next point and responds to the affirmative case and reasons
3rd Negative	closing remarks
3rd Affirmative	closing remarks
Judge	Decides who had the best arguments and rebuttal.

Emphasise Respectful Communication:

Establish clear ground rules for respectful communication, such as no interrupting, listening attentively, and avoiding personal attacks. Model respectful disagreement and encourage children to focus on the arguments, not the person making them. Explain that it's okay to disagree with someone's opinion, but it's important to do so respectfully.

Focus on Building Arguments:

Teach children how to find evidence to support their claims, even at a basic level. Encourage them to think about different perspectives and potential counterarguments. Practice brainstorming ideas and organising thoughts before speaking.

Make it Fun and Engaging:

Use games, role-playing, and other interactive activities to make the process enjoyable. Do it in the car or on long trips. This way it's a fun way to learn. Keep debates short and focused, especially at the beginning. Provide positive reinforcement and encouragement, even if the arguments aren't perfect. Celebrate effort and participation, rather than focusing solely on winning.

Practice Regularly:

Set up regular practice debates at home or in the car. Start with low-stakes topics and gradually increase the complexity. Encourage children to practice arguing both sides of a topic.

Debating in an AI World

Learning how to debate like this really helps to understand what is truth or untrue, what is possible or not possible and how things are different for different people. AI requires a real understanding of explaining what you mean and debating makes you explain your reasoning in detail. This is an invaluable skill for an AI world especially when you are learning to prompt AI for an outcome you need.

Fun Activities to support Debating

🍽 Dinner table debates: Vote on silly or practical topics before dessert.

- **Skills Built:** Verbal expression, active listening, reasoning, turn-taking, and confidence in sharing opinions.
- *AI-World Connection:* When families debate over dinner, whether about "Who does dishes better: humans or robots?" or "Should dessert come before vegetables?", children are learning to communicate with clarity and emotional control. These conversational skills prepare them for future roles in AI communication design, customer experience strategy, diplomacy, and leadership, where understanding nuance and perspective matters as much as logic.

 Car debate club: Pick a topic at the start of a drive and let everyone argue their case.

- **Skills Built:** Quick thinking, focus, and structured argument under time pressure.
- *AI-World Connection:* Debating on the go mimics the fast-paced decision-making children will face in AI-driven workplaces, where ideas must be articulated clearly and quickly. This practice nurtures flexibility, reasoning, and concise communication, vital for careers in AI

product management, entrepreneurship, journalism, and negotiation.

🔄 **Role reversal:** Ask children to argue the opposite of their true opinion.

- **Skills Built:** Empathy, perspective-taking, adaptability, and cognitive flexibility.
- *AI-World Connection:* Asking children to defend the opposite of their real opinion builds emotional intelligence and adaptability, two human advantages AI cannot duplicate. These skills prepare future policy makers, educators, mediators, and human–AI ethics specialists, who must see all sides of an argument before deciding responsibly.

🏆 **Debate nights:** Invite family or friends over for lighthearted "for vs against" challenges, complete with prizes.

- **Skills Built:** Public speaking, teamwork, humour, and persuasive communication.
- *AI-World Connection:* Hosting fun "for vs against" debates with friends or family helps children find their voice and learn how to present ideas with confidence and evidence. This confidence directly supports future success in media production, marketing, law, public relations, and leadership, where influence and storytelling shape global conversations alongside AI-driven media.

🔍 **Fact checkers:** Have children research simple evidence for their points—this helps them practise backing up claims.

- **Skills Built:** Research, data verification, critical thinking, and discernment.
- *AI-World Connection:* Fact-checking teaches children to verify infor-

mation before accepting it. This is an essential survival skill in the AI era. As generative tools create oceans of content, only those who can separate truth from fiction will lead wisely. This skill forms the foundation for future roles in data journalism, AI ethics, cybersecurity, academic research, and information verification.

AGE	ACTIVITY	EXAMPLE
1 – 5 years	Pretend Play	Build a Fort, play with a dolls house, make a cubby
	Dress ups	Pirate clothes, animal tails and ears, princ
	Why Game	Children get to say a statement and others ask Why? Then the child needs to explain and justify their statement
5 – 10 years	If I ruled the world	Children get to say what they would do if they ruled the world and then explain their reasoning.
	Convince you of something they want	Eg. Choosing a pet, having desert instead of dinner, they can have more screen time.
	Individual or group performance opportunities	Either with dance, an instrument, drama, competitions, gymnastics
10 + years	Debates in the car	Choose a topic such as would you prefer a dog or a cat for a pet and then take sides. Follow the rules of debating and discuss together.
	What would you bring if you were stranded on an island types of questions	Everyone needs to explain their reasoning behind their choices. You can then vote which choice you would bring as a family.
	Games	Cranium, Charades, Pictionary
	Board Games	Chameleon, Herd Mentality

COLLABORATION: WORKING TOGETHER FOR SUCCESS

Why Collaboration Matters

Collaboration doesn't come naturally to young children. Before the age of six, children are mostly egocentric. They're learning to share, but true collaboration is still beyond them. Once they begin to master sharing, they can also start to see the benefits of working together and collaborating. The quickest way for children to see the benefits of collaboration is to play together. Create a world together with lego, dolls, train sets, or any imaginative play toy. I used Little Pet Shop sets to create all sorts of streets and worlds for my children to play in. They had so many different sets they found lots of ways to creatively play with them.

> **Parenting Tip:** I found that the more I had of one toy, like lego or Little Pet Shop, the more my children wanted played with it and the more creative they got.

This type of creative play will help children to learn the collaboration skills they need for bigger projects with desired outcomes. As they move from collaboration during creative play, crafting, creative design to projects, the biggest lesson is how to make sure everyone is on the same page contributing with the specific skillset they have. Learning at home will help them to practise these skills so that when they have to do collaborative projects at school, uni and eventually the workplace, they know the benefits and pit falls. This way they can avoid doing 'all the heavy lifting' of the project and are less likely to be taken advantage of. You can teach this early by talking to them about the different roles people play in the collaboration space and start to give them techniques to spot those free loaders. Sometimes it means that you leave them to do most of the work and then talk through why that wasn't fair. Hands on practise is a perfect way to teach this to your children.

Creating a Collaborative Environment at Home

There are some collaboration techniques which are easy to implement in the home, such as cleaning and daily check-ins. Some other techniques will require help from others or are done during specific tasks such as making a flat pack together. It is important to use these collaboration skills throughout our lives to keep the skills in tip top shape.

Some simple practical ways we can weave collaboration into our daily family life are:

Group discussions:

Regular group discussions (family or friends), even for short periods, allow children to practice expressing their ideas and listening to others' perspectives.

Team-building activities:

Incorporate activities like building challenges or problem-solving tasks that require collaboration and communication to achieve a common goal. This is fun for the whole family. Building an IKEA item is the best team building home activity. My daughters and I had 20 cupboards to build when we moved into our new house a few years ago. We were able to make each cupboard in 20 minutes. Needless to say we were working as a well oiled machine in the end. Even to this day we can whip up a flat pack in no time.

- **Peer-to-peer learning:**
- Encourage your children to teach each other concepts or help each other with assignments. This promotes active listening, explanation skills, and a sense of shared responsibility.
- **Daily discussions:**
- At breakfast or dinner encourage students to share their thoughts and experiences, fostering a culture of open communication.
- **Create a supportive environment:**
- Emphasise the importance of respectful communication, active listening,

and constructive feedback to create a safe space for the family to learn from each other.

Explicitly Teaching Collaboration Skills

Here are some techniques which you will need to teach explicitly. By that I mean you teach through discussions and planned activities. Here is an overview of the types of skills you should teach in this way;

What is teamwork? Clearly explain what collaboration means and the behaviours that contribute to effective teamwork, such as active listening, sharing ideas, and respecting different viewpoints.

Conflict resolution: Teach your children how to identify and address conflicts constructively, focusing on finding solutions that work for everyone. Teach them through voicing how you are thinking when confronted with a conflict after the event.

Roles and responsibilities: When working in groups, assign specific roles with defined responsibilities to ensure everyone contributes and understands their part in the overall task.

Feedback and reflection: Encourage your children to reflect on any collaborative activity, whether that is cleaning the house or playing games. Make sure they can explain their own contributions, what happened and the families process and progress. Having them involved in the reflection helps them to understand how to solve the problems they experienced so it doesn't happen again.

Jigsaw strategy: Divide the family into groups to learn different parts of a topic, then have them regroup and teach their knowledge to others. This promotes interdependence and shared learning. As they get older you can have PowerPoint nights where everyone makes a slide show on different parts of a topic. Everyone learns and it is lots of fun.

Collaboration in the Age of AI

Collaboration is a real skill which can really take your children to the next level. Being respectful of others and learning how to all participate in an activity evenly can make the difference between good marks and average marks at school and university.

With AI doing the heavy lifting, it will be collaborations of human minds which ultimately determines the direction of companies, research and society. Collaboration isn't just a "soft skill" anymore. It's a critical workplace ability of the future. As artificial intelligence evolves, it will be human collaboration that shapes businesses, research and society as a whole. You will see more people working together, combining perspectives, and making decisions in business to guide our society into the future

Learning how to listen, respect others, and contribute evenly to shared projects will set your children apart. These are the very skills that turn good students into great leaders, and average teams into outstanding ones.

Fun Activities to support Collaboration

Family project nights: Build something together like crafts, furniture, or even a backyard fort.

- **Skills Built:** Collaboration, planning, patience, design thinking, and practical problem-solving.
- *AI-World Connection:* When families build together, whether it's a birdhouse, a fort, or a homemade bookshelf, children learn how to share ideas, divide tasks, and troubleshoot together. These are the same teamwork dynamics that power AI-era workplaces, where engineers, designers, and data scientists collaborate on shared goals. This activity

lays the groundwork for future roles in STEM innovation, architecture, AI project management, and creative entrepreneurship, where success depends on collective effort and flexible thinking.

Cooking as a team: Assign roles (chef, assistant, taste-tester, table setter) so everyone contributes.

- **Skills Built:** Organisation, communication, responsibility, and interdependence.
- *AI-World Connection:* Cooking together is like running a project with moving parts including planning, timing, quality control, and teamwork. It teaches children how to coordinate and adapt under pressure, mirroring real-world collaboration in AI operations, hospitality technology, event management, and systems design. Learning to "hand off" tasks gracefully builds leadership and delegation skills vital for future human–AI collaboration.

Group storytelling: Take turns adding sentences to a story. This fosters creativity and shared ownership.

- **Skills Built:** Creativity, listening, adaptability, and co-creation.
- *AI-World Connection:* When children co-write stories, they're practising exactly what future innovators will do which is building ideas together, one line at a time. This activity strengthens imagination, narrative flow, and teamwork, preparing them for roles in creative technology, digital storytelling, gaming, and AI-assisted content design. It's also a gentle way to show how multiple human minds (and someday human + AI) can create something richer together.

Board games: Choose cooperative games where players work toward a common goal.

- **Skills Built:** Cooperation, strategy, patience, and emotional regulation.
- *AI-World Connection:* Cooperative board games teach children to share strategies, read others' intentions, and work toward common outcomes key abilities in tomorrow's data-driven industries, simulation design, collaborative robotics, and leadership roles. When AI takes over repetitive tasks, humans who can work cooperatively, think strategically, and stay calm under pressure will stand out.

Family challenges: Set a collective task (like cleaning a room in 10 minutes) and celebrate the success together.

- **Skills Built:** Team motivation, time management, shared accountability, and celebration of success.
- *AI-World Connection:* Setting shared challenges like cleaning a room in 10 minutes or finishing a creative task together which builds teamwork momentum. Children learn that productivity and success feel best when shared. This understanding prepares them for future work in project coordination, agile development, leadership coaching, and human–AI collaboration, where motivating others and managing deadlines are daily essentials.

AGE	ACTIVITY	EXAMPLE
1 – 5 years	Chores	Accomplish as a family together. Each person has a chore and you can do them all at the same time. This helps children to feel part of a team and not slacking off.
	TV Shows	Bluey Bear in the big blue house
5 – 10 years	Chores	Accomplish as a family together. Each person has a chore and you can do them all at the same time. This helps children to feel part of a team and not slacking off.
	Movie	Inside Out, Upward, Red,
	Disney	A lot of the animation has lessons about values, choices, and legends from many different cultures.
	Board Games	Hanabi, Pandemic, Forbidden Island, Mumma Mia, Labrinth, Tzuro
10 + years	As above	

🎢 BUILDING RESILIENCE

We hear this word bandied around everywhere but what is resilience really? In essence it is the way we deal with problems in our lives. Resilient people have positive thoughts, action plans and learned behaviours to apply when faced with difficult situations so that they can recover quickly. A difficult situation can be as little as not getting your way when you are 2, to a fire destroying your house when you are 42. These situations are vastly different, however, how you cope with them and how you control your emotions during and after the event, shows how resilient you are as a person. The more little challenges you give your child early on, the more practise they get at using the coping strategies and the more resilient they will become as teenagers and adults. By practising on smaller problems they are then able to apply these coping skills to bigger problems when the going gets tough later in life.

Children who become resilient are able to change their way of thinking and acting when faced with adversity. They thrive on change.They believe in their own ability to get through any situation no matter what. In this ever increasing world of change which AI has bought, that is a skill worth knowing. It will be, and actually is needed now, in every part of their daily lives. AI challenges us in every aspect of our lives, it has increased change exponentially. By having resilience and self-worth children will be able to move with the times and make the most out of the AI future world.

The Role of Resistance

In todays developed western countries we are seeing that many children face very little resistance in their daily lives. They have plenty of food, all the basic necessities, and often most of the luxuries they desire. This is a wonderful thing and something that shouldn't be taken for granted. There is a balance even adults must find in our socioeconomic situations where we can both enjoy our luxuries while also having respect for ourselves and others. Sometimes this point can get lost when wanting to give our children the things we never had. On the surface, this seems helpful. We can afford it so why can't our children benefit? But in reality, constantly smoothing the path for our children denies them the chance to develop resilience and respect for what we earn or are given. Children build self-worth through working for goals and knowing that they are resilient. There should always be a balance as too much of anything can sometimes be harmful.

So how do we find that balance? It's through letting our children experience problems and work through them. Providing children with basic necessities, and some luxuries, is an amazing privilege for a parent, but its important to remember the brain learns through problems and making mistakes. This can be achieved by either witnessing them or better still, experiencing them. Making the mistakes, practicing the skills, feeling those bad feelings and knowing how to process and deal with them is all part of teaching the brain how to live in our world.

Lessons are learned when we are faced with uncomfortable situations.

Trying to keep your children happy all the time doesn't do them justice and will wear you out very quickly. We want to support and nurture but by taking care of everything and almost sheltering children from the realities of life will not help them to live their future to the full. Understanding and knowing that it is "ok to be sad" that you didn't get a solo part or make the soccer team, but still keep working towards next year is so important. It is all a part of developing **Resilience**.

Resilience also helps us, as adults, to deal with what life has to throw our way. Our emotional intelligence gets tested on a daily basis when we are bringing up children. We have to bite our tongue many times and hold back laughter when they say funny little things. We know that they watch how we deal with all situations and they learn from that.

Children may not seem to want to be guided by us, but they desperately need our guidance so that they are not unintentionally put through an event which, in some cases, could be traumatic. We absolutely still need to protect them with our guidance, but a little resistance on our part, like saying "no" to things, can go a long way. By making their path too easy in the beginning, you make their future harder. It is easier to learn new concepts at a younger age, when the brain is so malleable and tuned into learning.

Parenting Tip: It can be as easy as making children earn toys. Instead of buying the newest game immediately, let your child see the price, set a goal and ear it through practise, homework or extra chores. (Again, I believe that chores should not be rewarded by money as the family is a team working together for the good of all and this concept is very important throughout their lives.)

Yes there may be tantrums, but learning to process these emotions is part of building resilience. It shows them how the world works and increases their emotional intelligence by learning to move through the negative emotions of not getting the toy straight away, to working towards getting the toy and being more positive about the situation. The reward is a child who is grateful for what they get, has patience and learns to persist with a task to achieve a goal. All important lessons we use during adulthood.

The Ungrateful Entitled Beast

By giving them everything they want you are potentially unleashing an ungrateful entitled beast. If all your child knows is that they always get what they want, of course it would be distressing and confusing when life gives them a 'no'. People and situations in life are not always going to make your child's path smooth. There are going to be obstacles which need to be overcome and if your child has not been given these little resistant situations early on, they won't know how to cope with the challenges. This is when you will be called into the school to talk with the teacher. Not a fun trip for anyone. It's not about always saying "no" either. It's about balance.

Remember just because you can afford it, doesn't mean they should have it. Bill Gates' says his children aren't allowed technology until they are older, yet he is one of the biggest tech advocates. Of course he can afford it, but it's not about what you can afford, it's about what is best for your child's brain development. You saying 'no' is an important part of that development.

The Myer Tantrum Lesson

The hardest thing I ever did was stand in the front of a Myer in Adelaide with a 2 year old screaming at me that she "WAS NOT GOING TO GO INTO THAT SHOP". I had promised that after we bought a present for Nanna, we'd go to the restaurant for lunch. She was not having it. She wasn't tired, hungry, or sick. But she had made up her mind that under no circumstances did she want to go into the next shop. Some people would have just given in and gone straight to the restaurant, but I had stamina too. I knew that if I put in the effort now I was helping to train her brain and build resilience for her future life. Some people would say that maybe I was just as stubborn as she, however I was standing there, waiting for her to calm down, knowing that this was a line in the sand which had to be drawn.

I was quickly approached by one of the security guards asking what was happening. He had been called over to help by another person in the shopping centre. He was a big man with a warm smile. I saw him come over and was prepared to scoop her up in my arms and make a quick escape. I explained that she was having a tantrum and I will never forget his words of wisdom. *"Yes Mamme you just stay there. I have 10 children and this is normal.Don't you give in or it's a slippery slope. You're doing the right thing. I'll just stand here so no-one bothers you."*I remember thinking *"Wow 10 children, how does he do that when I can't even control this one"*, but then I took a breath, smiled at him and continued on dealing with my screaming toddler, with him standing guard. I didn't say anything to her except *"We will got into the shop and get Nanna's present then go to the restaurant. Let's go."* Repeatedly. Again and again I kept saying the same thing until after what seemed like 5 hours (which was probably only 5 minutes) she got up and walked into that shop.He nodded to me and walked away with a smile. His words of wisdom and support kept coming back to me over the years when I felt the challenge of parenthood and knew I had to stick to my guns for my children's sake. Know I know that this is how they learn.

Repetition and resistance builds resilience.

Previously in these situationsI was just flying by the seat of my pants not understanding that it was so important that I stuck to my guns. Now I understand that my children were learning how to deal with the emotions and building their resilience through the process, no matter how loud they were being. By building resilience and learning how to deal with the emotions which are involved in 'no' you are preparing them for the changing world which is AI.

Choosing when to add resistance

It's important to choose wisely when introducing resistance. Everyday boundaries can help to teach resilience. It might look like working towards toys, treats, or screen time. These are perfect opportunities to build resilience through resistance. There is a time, when I believe, that you can spoil your child and that is on holiday. It's the everyday indulgences that undermines resilience not the special experiences you have together on holiday.

Holidays provide children with many opportunities to experience change. A change in their bed routine, a change in their eating, a change in their environment. There is a lot of change there, but there is also a lot of wonderful experiences which will shape their memories and ability to understand the difference between AI fake or AI truth in the future. (We will go into this more in Intermediate Skills.) These new experiences on holidays are so important for their brain development that I advocate for spending as much time and money as you like on a holiday. These opportunities don't come often enough. You also need to 'fill your cup' and have a break. So spend up big on experiences, take lots of photos to remember the holiday and buy up special souvenirs to remember the fun you all had together. All of this knowledge will be used throughout their lives. They will have a new perspective on their own house, school and friendships. The holiday will help them to build their own view of the world and give them a basis to tell what is truth and what is not in the AI world they are about to enter. This is not a time to practise adding resistance when there is already so much change.

🔆 **Parenting Tip:** Spoiling your children is ok on holidays just not the everyday. It's the everyday times when resilience is learned through small situations you create, when everything else is stable and they have the best chance of learning how to deal with the emotions they are feeling.

Music as a Masterclass in Resilience

Another powerful way to build resilience is by learning an instrument. There are so many benefits of playing an instrument. It uses both hemisphere's of the brain concurrently. This is important as it; *Improves cognitive ability* (problem solving, memory and focus). *Enhances complex tasks* such as processing languages and solving complex problems. *Boosts physical co-ordination* as you need to use both hemispheres to move each side of your body for gross motor skills like swimming or dancing as well as fine motor skills such as writing. *Promotes overall performance* in any field as the synergy between the two hemispheres is paramount to combining creativity and imagination (right hemisphere) with analytical skills and formulating hypotheses (left hemisphere).

It also helps to develop skills which they will use in their studies such as concentration, impulse control, working memory as well as short-term and long-term memory. It doesn't stop there, they learn to stay focused, delay gratification (which is difficult in this digital world of instant reward and gratification), learn how to correct mistakes, and take feedback with no emotion attached, accept feed back constructively and how to stick with practice even when they don't feel like it to reach a goal.

The brains ability to develop neuroplasticity is almost super tracked by music as it uses both hemispheres and has so much repetition built into the process of learning. This doesn't come from just playing once a week though. It is the everyday practise which develops these benefits to the children's brain and life skills. It is them having a controlled predictable resistance in their lives everyday which rewires their brains.Over time children can see how consistent effort leads to mastery.

The Lightbulb Moment

When my daughter was in year 2 she was finding reading really difficult. She had been playing the flute for a few years everyday and had seen the progress she was making and the difference everyday practise had on this progress. Everyday after school I would sit down and have a cuppa while they unpacked their school bags and got into their home clothes. I found that when I just sat at this time they would come up and tell me about their day, but if I was busy cooking dinner or cleaning, they would just play. But I digress. She suddenly ran into the living room. "Mum, Mum. Do you think if I read everyday it will get easier, just like the flute?" This was her lightbulb moment and it was a big one for her. People often don't get this concept until much later in their school life and she was in year 2. All the work I had put in giving her incentives to keep practicing everyday, all the money I had spent on these incentives were all worth it. Once learned, she then went on to apply this new concept to all of her school work and low and behold, schoolwork became easier for her.

Building Mastery

Building mastery of anything can give your children the advantage when they are adults. Malcolm Gladwell writes about the 10,000 hour rule. I didn't really believe it until I saw it in action with my daughters. One did it with music and the other with dance. Both started their discipline at about 3 years of age. They practised everyday and when opportunities presented themselves they went for it. When the flute player started playing in the senior orchestra when she was in year 5 she was also asked to play in the jazz band and the junior orchestra. She went from her lesson (½ hour per week) and practising everyday (3.5 hours per week) to adding another 6 hours of playing per week. Then she was asked to play for their musical which was another 3 hours per week of playing and it then snowballed with special performances and State Youth Orchestras. She was celebrated and having so much fun. Even today she tells me she enjoyed all the orchestras she played in most. Her time playing the instrument went from 4 hours per week to 16-20 hours per week. The numbers quickly add up when you do this from year 5 to 12. It is about 6,500 hours for just the basics.

This is exactly what was happening to the ice hockey players who Malcolm Gladwell writes about. The more they played the better they got and the more opportunities came their way. This is how we build mastery. Understanding how to master a skill can then be applied to any skill. So many successful olympians are successful in other ventures in their lives, either with study or business. You will also find successful professional sportsman often become successful business men in their own right after they retire from sport. Teaching your child how to master a skill early on will help them to reach their potential in life.

It's not just the mastery which helps them in life. Learning to practise through the tough times when they don't want to, but have to, is paramount to living in our society today. There are so many times I hear people saying "Why do I have to cook again?" We know we have to but it get's a bit monotonous.

Having worked through the emotions early in life, you can just put on a smile and do it without any drain.

You can be a positive influence in peoples lives. It's having those human connections which make us different to AI. Knowing that the hard times will end and the fun will come back is priceless. These are the skills mastering something gives you. It might take a few years for them to understand this but once they do, it helps with so many other parts of their lives. The best part is that it helps children to understand the process of learning which means that they can learn easier at school, university and throughout their working life.

Fun Activities to support Resilience.

Chore challenges: Set small goals and reward effort, not just results.

- *Skills Built:* Responsibility, consistency, intrinsic motivation, and delayed gratification.
- *AI-World Connection:* When children learn to complete small, routine tasks and celebrate effort, not perfection, they're developing the mindset that fuels success in an AI-driven world. These micro-habits mirror the perseverance needed for coding, robotics testing, and long-term

creative projects. Future roles in engineering, research, healthcare, and entrepreneurship all require steady, process-oriented thinkers who stay motivated even when results aren't immediate.

⏳ **Waiting games:** Practise patience by timing short waits and celebrating success.

- *Skills Built:* Patience, emotional regulation, focus, and impulse control.
- *AI-World Connection:* In a world of instant answers and one-click gratification, patience is a rare superpower. Timing short waits and celebrating success helps children strengthen the neural pathways for focus and emotional control which are skills that AI can't replicate. These qualities are vital for future scientists, analysts, surgeons, programmers, and leaders, who must think long-term, tolerate ambiguity, and stay calm under pressure.

⚽ **Sports or hobbies:** Encourage persistence through setbacks and learning curves.

- *Skills Built:* Perseverance, resilience, teamwork, and growth mindset.
- *AI-World Connection:* Every missed goal, dropped catch, or imperfect routine is a real-world lesson in persistence. Sports and hobbies build resilience through repetition and adaptation which are qualities essential for tomorrow's AI engineers, designers, innovators, and data scientists, who must test, fail, and refine constantly. Learning to push through setbacks creates thinkers who thrive in iterative, experimental environments.

🎼 **Music practice:** Use consistent routines to build discipline and mastery.

- *Skills Built:* Discipline, concentration, goal-setting, and long-term focus.
- *AI-World Connection:* Regular practice trains the brain for mastery,

structure, and creativity. This is the same balance required to code, compose, or design alongside AI. Musicians learn pattern recognition and attention to nuance, foundational skills for AI algorithm design, digital media composition, cognitive science, and creative technology. It's where art meets structure, the perfect preparation for the AI era.

"No" days: Occasionally say no to requests and guide kids through the disappointment.

- **Skills Built:** Emotional resilience, adaptability, self-regulation, and perspective-taking.
- **AI-World Connection:** Occasionally saying "no" and guiding children through the resulting frustration helps them develop emotional intelligence and coping skills that AI will never understand. In an automated world where many desires are instantly met, the ability to handle disappointment with grace becomes a true human strength. This emotional resilience is crucial for future careers in leadership, counselling, education, design, and AI ethics, where empathy and perspective guide better decisions.

⛰ GRIT: CLIMBING THE MOUNTAIN OF LEARNING

Grit is the ability to push through the hard, uncomfortable phase of learning and come out the other side with a new skill. It's sticking with something long enough to achieve mastery, even when it feels difficult, frustrating, or slow. It is part of the learning process for all our learning.

Grit is a term created by Angela Duckworth. She uses this term to describe the ability to push through the hard part of learning, to end up with a new skill. Children start to learn about Grit from the very beginning when they start to roll over or learn to walk. You never see a baby try to roll once and then say "OK, I can't do that, what's next?" No, they try and try and try and once they have mastered the skill, then they do it a thousand times.

When one of my girls was in grade 2, she was finding it really hard to learn to read. It was a real struggle. As a background I need to tell you that she had been learning the flute for 3 years by this stage and we had pushed through many pieces which seemed nearly impossible at the start, but with breaking them down to little parts and playing them many times, each repetition became easier. Once the piece was easily played, you didn't just put it away, our teacher asked us to continue to play it everyday so you can master the skill which each piece teaches. Every piece, every day was the philosophy and it worked really well as she loved feeling the success of playing pieces easily. This is in essence the philosophy of Grit. Work hard on a skill, push through the repetition, until it is mastered. Many times she would find a new piece "too hard", only for it to become her favourite piece. Anyway, we were on our way home from school one day and she was complaining of getting 'another book' from her teacher which she was suppose to read, then the lightbulb moment hit her. She paused mid sentence and I will never forget these words "Hey Mum, if I read everyday like I practice my flute, maybe I will find reading easier? Do you think that would happen?". "Absolutely" I replied with a sigh of relief. She had made the connection of learning through Grit and from then on, she applied herself to reading just like she had applied herself to the flute. Within a few months she was reading Harry Potter and loving it.

It's not about getting it right straight away. You might get it wrong and make mistakes at first, but you keep going. It's the 'Not Yet' philosophy. This is what we want our children to do with all learning experiences. Not to give up because it is too hard. Children will start to understand that they can learn anything with Grit. It is worth checking out Angela Duckworth's books, her latest is definitely worth a read.

Grit applies to every area of life:

- **Physical skills** – learning to ride a bike, play an instrument, or write your name for the first time.
- **Academic skills** – learning to read, solve maths problems, or master a new concept.
- **Emotional skills** – learning how to lose gracefully, persevere after failure, or regulate emotions in challenging situations.

The Grit Process Described

In the picture above you can see that the goal is the skill at the end of the mountain where the finish line is located at the fireworks.We start at the beginning all confident and ready to accept the challenge. Then very quickly we dip way down into the pit.

This pit is where we start to learn the new skill. You can see the long way we have to go before we can do the skill. The road ahead is hard and long but it gets short and shorter. If you have ever climbed a steep mountain you can relate to how hard that climb is and how wonderful it is to look into the distance once you are on top of the mountain. This is the feeling you get when you can do the skill. It is important to celebrate when you reach the top, just like you will stop sit and have a break when you reach the top of a mountain climb.

While children are in the process of learning a skill, you can ask them where they think they are on the learning journey. They can visually see that it is hard work to acquire the skill. It doesn't matter what you are learning, it is all the same. It takes grit and determination to get to the top of the mountain. What a wonderful analogy to help children to understand the learning process.

Teaching the Mountain Analogy

Imagine learning as a mountain climb. At the top of the mountain is the goal: the new skill you want to master. Fireworks burst at the peak, celebrating your achievement.

- The Starting Point: At the base of the mountain, children often feel excited and confident. They're eager to begin the journey.
- The Dip into the Pit: Very quickly, the enthusiasm fades as the climb gets tough. This "pit" is where true learning begins. Mistakes, frustrations, and struggles set in. The road ahead looks long and daunting.
- The Hard Climb: Step by step, progress is made. It feels slow, but gradually the path shortens and the summit comes into view.
- Reaching Mastery: At the top, children look back and see how far they've come. They feel the pride of perseverance and celebrate the fireworks of success.

This analogy helps children see that learning always involves struggle—and that the struggle is a normal, necessary part of the process.

꙾ Parenting Tips:

Use the Mountain Visual: Show children a picture of the mountain and ask, *"Where do you think you are on your learning journey?"* This helps them connect emotionally with the process.

Normalise the Pit: Remind them that the dip into the pit isn't failure—it's where growth happens.

Celebrate Persistence, not just results: Praise their effort in climbing, even before they reach the top.

Link it to Real Experiences: Share your own stories of grit—whether learning a skill as a child or persevering as an adult.

Why Grit Matters in an AI World

AI can process information instantly. It can be used to find out and process data as well as analyse information, but it cannot help children to develop grit. Perseverance, resilience, and determination are uniquely human traits. By teaching children grit, we prepare them to face challenges that no machine can overcome for them. Grit is the muscle that powers lifelong learning.

Fun Activities to support Grit.

AGE	ACTIVITY	EXAMPLE
1 – 5 years	Learning an instrument	Suzuki Music Methodology is best for under 5 years
	Learning a sport	Little Athletics, Soccer, Basketball, Tennis, Swimming
5 – 10 years	Learning a physical skill	Ballet / Gymnastics / Circus / Calisthenics
	Games	Long card or board will give you practice in many emotions and keeping on going until the end. primarily ones which take time such as monopoly, cluedo,
	Puzzles - 300 pieces or more	300 + pieces
	Crafting	Beads, Paper, Recycling construction
	Physical challenges	Hula hoop, Juggling, Skipping rope
	Building challenge with lego	Make something without specific parts or instructions
	Origami	
10 + years	All of the 5 - 10 years	
	Outdoor challenges	Skiing, water sports, Long treks / bush walks
	Drawing challenges	Art Hub do great drawing instructions but then try to create your own drawings
	Crafting – learning a new skill	Making something like a pillow with a sewing machine, or knitting a pillow, or making a soft toy which you have never done.

⚜ ADAPTABILITY: THRIVING THROUGH CHANGE

What do we mean by being adaptability? Adaptability is the ability to adjust to new situations with ease rather than frustration. Put simply, it's the difference between calmly shifting gears and having a meltdown when plans change. It's about taking the emotion out of the situation and deal with the actions. Some children pick up these skills naturally. Others need explicit teaching.

Children learn to be adaptable by experiencing a change of plan or losing a game. Have you seen those people who lose a game and flip the table. They rant and rave and it makes it very uncomfortable for everyone. They just don't know how to manage those big emotions which are stirred up. Learning to manage those feelings is part of becoming adaptable. One of my daughters would never be able to walk away from a situation where she didn't believe that the other person agreed with her. We would be forever saying "Drop the bone. It's not worth it". It took her many years to be able to "drop the bone". It was hard for her to move through the change, but now she is able to thrive in change. Most ironically, her job requires her to guide others through change.

The future AI world will required workers to be able to use AI to get their work done. AI changes so quickly that those people who are able to move with the changes AI is bringing and adapt to the new technology will be front runners in the job market. Adaptability is all about how people can thrive in the new or changing circumstances where as resilience and emotional intelligence is all about the ability to understand, express and manage your emotions. You need these two skills to be able to become adaptable.

Resilience + Emotional Intelligence + Problem Solving = Adaptability

Some examples of being adaptable are;

- **Handling routine changes:** Being okay when bedtime is later than usual, or when a different parent picks them up from school.
- **Shifting strategies:** If a puzzle piece doesn't fit, they try a new way instead of giving up.
- **Accepting mistakes:** If you lose a game, you are able to bounce back, instead of melting down.
- **Making new friends:** Joining in with a group of children they haven't played with before.
- **Problem-solving on the fly**: If the park is closed, being able to enjoy a different activity instead.

- **Managing transitions:** Moving from home to school, class to recess, or primary to high school with resilience.

Teaching Adaptability Through Narrative

One of the most important ways to teach adaptability is through narrative. That is, talking through your thought process when changes happen. The children then start to understand what is happening in their own bodies and mind and how it should look as they listen and watch you go through the process. We are not saying it is easy, but it is doable.

Children learn through watching and listening to you. A simple phrase like *"Oh, it's raining, so we can't go to the park. Let's think of a fun indoor game instead."* Starts the process. They will then watch what you do. If you find a new activity and have fun, then they are more likely to roll with the changes and have fun too. If you are in a bad mood for the afternoon, complaining how awful it was that you couldn't go to the park, then they are more likely to kick up a stink when you ask them to change their actions or minds on a topic later. They learn through watching how you react to the situation. A **<u>positive attitude</u>** on your part, goes along way to developing a flexible way of thinking.

By talking through the situation you are letting them know what is acceptable behaviour and what they should be doing. They will follow your lead in every situation. They remember what is important to you and will do this every time. They generally follow actions more than words.

Practical Ways to Teach Adaptability

Here are some quick ways you can teach your children how to become adaptable.

Model Flexible Thinking - Talk through your options and be positive about the new outcomes.

Introduce Small, Safe Changes - Mix up routines in a small way.

> 💡Parenting Tip: I always use to frame the issue as an adventure so that they looked on it as a fun exercise rather than something to worry about. When our house was destroyed by a hail storm and we had to go to a motel room for a few weeks, we had all sorts of "adventures" together, like seeing how many shops within walking distance of our room sold chocolate. It gave us exercise so we weren't cooped up in a room and we had a treat as well at the end. A win win situation the girls thought.

Teach Problem-Solving Skills - While driving or sitting somewhere talk about 'what-if' scenarios. "What would you do if your friend wouldn't let you play today?" "What would you do if alien came down and took our dog?" They can then brainstorm as many solutions with explanations as possible. It doesn't need to be a real scenario, but it helps them to go through the problem solving process to get the hang of it. Then, when a real situation happens they have a number of problem solving techniques they can use.

Build Emotional Regulation - We have spoken about Emotional Intelligence.This is an important part of that. It includes helping them to have positive self talk and knowing when to exit a situation to calm themselves down. Adaptability grows when children can manage big emotions.

Encourage Growth Mindset - A growth mindset will allow your children to get through tough situations. It is important to note that praising effort and not just the outcomes and reminding them that mistakes are all part of the learning. The power of perseverance is an important part of a growth mindset along with Grit.

Play Games That Require Rule Flexibility - Uno and Jenga are great games where there are different rules for everyone. You need to agree on the rules in the beginning of the play. Depending on who is playing they may be different for each person. Once you have agreed on the rules, then you must follow them all the way through the game. Simon Says is another game which is great to teach about flexibility, as the children need to make sure they listen to 'Simon' and the game changes constantly.

Celebrate Adaptable Moments - This can be just a quick comment when something changes and they move on with no issues. "You were really flexible when John had to leave early. I noticed you found another way to have fun with new friends." It's all about reinforcing the skill to make it stick. Games can really help, but just a quick comment is enough.

Fun Activities to support Adaptability.

ꟿ **Switch It Up:** Change dinner seating or bedtime routines occasionally and frame it as a family adventure.

- **Skills Built:** Adaptability, openness to change, and positive reframing.
- *AI-World Connection:* When families switch dinner seats or bedtime routines and call it an adventure, children learn that change isn't scary, it's an opportunity. In a world where technology evolves weekly, flexible thinkers thrive. This small family exercise nurtures comfort with uncertainty, preparing children for future roles in innovation, product design, AI development, and leadership, where agility and optimism drive progress.

ꟿ **Obstacle Course:** Create one indoors or outside that requires different strategies each round.

- **Skills Built:** Problem-solving, strategy, resilience, and coordination.
- *AI-World Connection:* Navigating an obstacle course especially when

the rules or layout change. Being able to change with the rules teaches children to plan, adapt, and think on their feet. These same cognitive skills underpin success in robotics, engineering, emergency response, and AI system design, where rapid adjustments and creative troubleshooting are part of daily life. Every new obstacle is a mini-lesson in agility under pressure.

📖 **Story Switch:** Start a story and have each family member change the plot mid-way.

- **Skills Built:** Creativity, perspective-taking, and collaborative thinking.
- *AI-World Connection:* Passing a story around and letting each family member twist the plot mid-way encourages flexible thinking and co-creation, just like working in human–AI creative teams. This game builds imagination, improvisation, and openness to new ideas. These skills are essential for future careers in creative writing, gaming, interactive media, and human–AI storytelling design.

🎲 **Rule Changers:** Play board games with modified rules to practise flexibility.

- **Skills Built:** Adaptability, reasoning, fairness, and flexible strategy.
- *AI-World Connection:* Altering board game rules helps children practise fairness while adjusting to new systems. Being able to adjust is exactly the mindset needed in a world of evolving technologies and shifting algorithms. It nurtures resilience and cognitive flexibility, preparing them for fields such as AI policy, strategic leadership, law, and software design, where innovation often means rewriting the rules.

🎁 **Surprise Days:** Plan a small change in routine and let children practise responding positively.

- **Skills Built:** Emotional regulation, positive mindset, and curiosity.
- *AI-World Connection:* Planning small surprises, like breakfast for dinner or an unplanned family outing, teaches children that unexpected changes can be fun, not frightening. This helps them build the calm curiosity needed to thrive in unpredictable environments such as start-ups, space exploration, scientific research, and AI innovation, where flexibility and emotional intelligence are invaluable.

AGE	ACTIVITY	EXAMPLE
1 – 5 years		
5 – 10 years	Board Games	Story-time Chess Jenga Pictionary
	Puzzles	Sudoku Crosswords
	Stories	Story chain adventures (one person starts and then the next continues
10 + years	Challenges	Escape rooms Scavenger hunts Teamwork games Charades and Role-playing games Pictionary Rubik's cube - solve in a time limit.

FINAL THOUGHTS ON FOUNDATION SKILLS

This foundational chapter lays the groundwork for everything that follows in *The Advantage Code*. It begins by introducing the essential early skills such as emotional intelligence, social and living skills, collaboration, debating, resilience, grit, and adaptability.These form the backbone of every child's development. These are not just "nice-to-have" character traits; they are the human advantage in an increasingly automated world.

From a toddler's first tantrum to a teenager's emotional storm, children are learning how to recognise, regulate, and communicate their feelings. Through consistent guidance, structure, and repetition, parents can nurture empathy, emotional control, confidence, and problem-solving — the invisible toolkit that allows a child to thrive both online and offline.

Core Skills and Why They Matter

❤ *Emotional Intelligence (EQ):*

The heart of this chapter lies in EQ. It is the ability to recognise, express, and manage emotions while understanding those of others. As AI takes on logical, repetitive, and analytical tasks, it is emotional intelligence that will separate human talent from machine performance. Empathy, compassion, and kindness which are the very traits AI cannot replicate, will underpin careers in leadership, education, mental health, customer experience, human resources, and diplomacy.

Social and Living Skills:

Simple acts like greeting others kindly, taking turns, or helping with household chores build cooperation, empathy, and community-mindedness. These daily habits translate into workplace collaboration, respect for diversity, and responsibility which are all vital in the team-driven environments of the future. Whether your child becomes a nurse, designer, engineer, or project manager, success will depend on how well they can work with and for people.

⚖ *Communication and Debating:*

Debating teaches children to think critically, reason logically, and present their ideas clearly. The foundation of creative problem-solving and truth-seeking. In an era of misinformation and AI-generated content, children who can analyse evidence and express persuasive arguments will excel in roles like data ethics officers, policymakers, journalists, lawyers, and innovators guiding AI use responsibly.

🤝 *Collaboration:*

Teamwork transforms individual intelligence into collective progress. Children who can listen, share, and compromise will be the change makers of tomorrow. From research teams designing ethical technology to creative studios developing human-AI partnerships collaboration will enhance their work results. As automation reshapes industries, collaborative skills will remain timelessly human.

⛰ *Resilience and Grit:*

Resilience and grit are the muscles of learning. They help children push through frustration, recover from failure, and keep climbing the "mountain of mastery." In the fast-moving AI economy, where job roles evolve faster than ever before, those who can learn, unlearn, and relearn will lead. These are the future entrepreneurs, problem-solvers, engineers, and creators who adapt instead of retreat.

↻ *Adaptability:*

Adaptability turns uncertainty into opportunity. It's the bridge between emotional strength and practical problem-solving. In the AI world, where tools and systems change overnight, flexible thinkers will thrive in all areas of industry. Whether designing sustainable cities, managing global teams, or working in emerging fields like human-AI collaboration, digital design, and virtual education adaptability will be the defining feature of sustainable careers.

As we have talked about before. To stay relevant these skills will help your children have the human connection. An AI-driven customer service or leadership role will require people to understand emotions and create an empathy connection where the automation cannot. Workers in all areas need to be able to navigate deepfakes, misinformation and data bias which is created by AI. To do this they will need critical thinking and augmentation skills which our foundation skills help to provide. The future careers won't be linear. People will need to upgrade their skills continuously to remain a valuable member of the team. This is where resilience, grit, adaptability and life long learning is essential to make your way through the transformation of industries which AI is creating. Being able to collaborate with others and technology as well as pivot ideas to stay relevant and fulfilled in the new future AI workplace will give your children the edge they need to succeed.

Children who master these foundational traits are building readiness for:

- Human-centric roles: educators, healthcare workers, counsellors, community leaders.
- Creative and design fields: artists, writers, content creators, UX/UI designers, marketers.
- Ethical and leadership pathways: AI ethicists, policy advisors, diversity and inclusion officers.
- Entrepreneurial innovation: founders who lead AI-integrated start-ups that prioritise social good.

These jobs of tomorrow demand empathy, emotional regulation, communication, and collaboration which are the very qualities this chapter teaches.

Final Thoughts

This chapter reminds us that before we teach coding, data analysis, or robotics, we must first teach kindness, patience, and perseverance. These foundation skills are the roots from which every other talent will grow. In an AI-shaped world, humanity is the ultimate advantage — and it begins here, in the conversations, boundaries, routines, and repeated small lessons of everyday family life.

FOUNDATION SKILLS AT A GLANCE

Building human advantage in an AI world

EMOTIONAL INTELLIGENCE

Future Jobs: Leaders, Teachers, Healthcare Workers, Psychologists, Customer Experiance Experts

AI World Advantage: empathy, relationship - building and emotional awareness remain uniquley human and irrepleceable.

SOCIAL & COMMUNICATION SKILLS

Future Jobs: Media Presenters, Team Managers, Human Resources, Diplomats, Marketing & PR

AI World Advantage: Translating emotion into action, The ability to inspire, mediate and connect across cultures and platforms.

DEBATING & CRITICAL THINKING

Future Jobs: Lawyers, Policy Advisors, Data Analysts, Researchers, AI Ethica Officers

AI World Advantage: Distinguishing fact from fiction, Reasoning through complexity and making ethical decisions.

COLLABORATION

Future Jobs: Engineers, Scientists, Designers, Product Developers, Entrepreneurs

AI World Advantage: Merging human creativity with technological precision - Teamwork that drives innovation.

RESILIENCE & GRIT

Future Jobs: Innovators, Founders, Athletes, Artists, Long term researchers

AI World Advantage: Perseverance through rapid change and uncertainty, embracing lifelong learning.

ROOTS: CORE FAMILY PRACTICES

Consistency - Boundaries - Repetition - Conversations - Experience

4

INTERMEDIATE SKILLS IN DEPTH

<u>**Suggestion: 5 years +**</u>

Now that we've built a solid foundation, it's time to take things up a notch. These **intermediate skills** are where curiosity starts to meet critical thinking, and where your child learns how to make sense of the complex, information-filled world around them. Skills I focus on are data protection/v erification, general knowledge, research skills, how things work and problem solving. These skills are more advanced and build from the foundations. They help children not only **use AI wisely**, but also **question it, guide it, and improve upon it** which will be paramount in the future AI world.

As before, you'll find **stories, practical tips, and end-of-chapter game tables** to help you bring these lessons into your family's daily life. Pick and choose the strategies that fit your household, and introduce them at your child's pace.

Real learning happens through practice, reflection, and repetition.

These skills won't develop overnight, but the more opportunities your child has to test, question, and explore, the stronger and more adaptable they will become. So, let's dive into the next stage of their journey:

Turning curiosity into capability.

▨ DATA PROTECTION AND VERIFICATION

Teaching children to Tell Fact from Fake

In an AI-saturated world, telling **truth from fake** gets harder every month. Children need real-life experiences and simple routines to ask, *"Does this make sense?"* and then **verify** what they see—whether it's a photo, a claim on social media, or a too-good-to-be-true "fact." It's not about memorising every detail; it's about **thinking like a checker**. They need to be able to analyse what is being given and be able to verify the data. Making sure that they ask the question "Does this make sense?" when they see any image or read any article. Is the response AI is giving a truth or just a fabrication.

If we ask AI where milk comes from you might get an answer - from the shops, or - from a cow at a farm. Which one is correct? In actuality they both are correct. We can help our children to analyse the outcome by talking through some of our thought processes. If they had never been to a farm or heard that milk comes from a cow, but go with Mum and Dad to the shops to get the milk, then the second answer (from a cow) would seem like a fabrication, an untruth. It's about learning how to understand the steps of how to critically assess a situation with your general knowledge of the topic and a science understanding of how the world works.

I remember my dad talking about a research paper he was looking at. He was indignant that these researchers had used false data in their paper to prove their point and this was over 20 years ago, well before AI.He pulled out the research paper they had quoted and showed me that the sample size of participants was below 20 a very small sample size. He was going to make sure I knew how to verify information. He showed me that it even said in the paper that it was too small of a sample size for them to come to a definitive conclusion, however the numbers suited the initial research paper so they quoted it and used it as a backup for their arguments. As a result of this, the paper could not be used for referencing in his job.

It is this fact checking which we need to ensure happens with AI. AI only is as good as it's prompts and the most common information held true. When using AI it is imperative that you fact check. There has been news of a company who has had to refunded hundreds of thousands of dollars because of false data being presented in a report which was collaboratively created by humans and AI. We need to be able to ask the questions; Is this true or not? and How can we tell?

AI is only as good as it's data

AI can only use the data it's given and extrapolate from there. If that data is lacking or incorrect or the prompt is not encompassing of enough details, then it can give us incorrect answers. It uses the most popular data, which isn't always correct, or properly researched. AI is intelligent enough though to keep asking questions. I'm sure you agree that talking to AI programs is awe inspiring in what they can do. Beware. AI is evolving very quickly.

Currently AI searches are based on previously known and collected data on the internet. To make the most out of AI programs, we need 'clean' data. What I mean by 'clean' data is information which is researched and not just the opinions of others. Data which is collected and verified as correct. It is important that AI has the information/data which is appropriate to use so that AI can give us relevant information and not just the most popular ideas.

JP Morgan's Secret Weapon

JP Morgan knew the value of this. They set up their own AI system for finance only using their own data sets, not all the information on the net. The gathered the information from hundreds of educated employees (who have an enormous amount of experience) and asked them to input their knowledge, their own understanding of the market, past trends and what their future predictions were into their own AI information system. The AI system was then asked to use this specific data, analyse it and voila, they had created their own data sets. These closed data sets, which are specifically "JP Morgan's brain power" gave them the edge over other big financial companies who were using information which was the opinions of uneducated AI. They used their own information, not the vast amount of data on the internet which could not be trusted. Why do they do this you say? This way they avoid data which is contaminated with incorrect information. It makes the AI system much more powerful and trustworthy.

So the question is; How do we do this as parents in our daily lives to help our children get the same result?

AI is changing what we need to learn

We also need to consider that AI is changing the way we learn and what we learn. For example when calculators came in, we didn't need to know exactly how to do the calculation of percentage but we did need to know it exists, why we need it, and how to use it. The mechanics were done for us on the calculator. We still needed to know about numbers and how to use them, we just didn't need to remember the exact formula. This is much like AI.

Spellcheck

This is where computing systems have helped but also sometimes hindered us. I'm sure you've all had spell check change what you've typed and been horrified at the result. Spent a good 10 minutes laughing and then another few minutes trying to resolve the mistake. Spell check was created because people find spelling hard. We intrinsically know when a word is spelled right, or not, depending on our experiences. We don't need to know what the actual spelling is, just that; the word exists, what it sounds like, when it's right and most importantly where to verify the spelling.

We now need to do this with AI. AI is a knowledge base of information and answers on steroids for every aspect of our lives, processes, businesses, marketing etc. Knowing when things are right or not is the key to being able to be successful in an AI world.

There is an old saying "You don't know what you don't know."It sounds weird but if you think about it, when you buy milk from a container all your life, how can you know that it comes from cows and a farm if you haven't learnt that, seen that or been told? In essence

"You don't know what you don't know."

What is "data" (and why should children care)?

First of all children of this age need to understand what **data** is. It is not only information. It is photos, words, documents, processes and websites. By the age intermediate skills should be taught children will be using technology more and more throughout their lives. Maybe you have given them a game console, they use your phone or they just got their own phone? They will be using iPads or computers in school, learning how to use technology to complete assignments or projects. Recognising what produces data and how to manage it will be important first steps in the skills of data protection and verification.

The essential skills which children need to know with relation to data is:

- How to **create** data (a photo, a doc, a video).
- How to **protect** data (so it isn't lost, stolen, or altered).
- How to **verify** data (so they're not fooled by fakes).
- How to **analyse** AI's output against real-world knowledge.

Creating and Managing Data

Children need to be guided in the importance of saving their data so they don't lose it. Using relevant file names and creating folders for data storage is a skill which parents can teach at home. The most important aspect of managing their data will be knowing how to protect your own data from corruption and or theft and should go hand-in-hand with learning how to produce your own data.

Protecting your data, such as your image or voice, from corruption or AI has become a real issue. In Denmark they have new copyright laws which protect your face and voice from replication by AI. This is a landmark law and one which will be followed by many others with relation to AI in the future I'm sure. The best way to teach children about data protection is to explain that they are like your jewellery or valuables. Keep them close and monitor them constantly.

Data verification and data analysis. Data verification and data analysis is another skill which is needed in the AI world. It forms the basis of protecting your data. AI is capable of running a data verification or data analysis but it all comes down to your prompt and AI understanding how the real world works, which will be the game changer if they can do this in the future.

So how to we give our children the skills to be able to protect and verify AI data. It might sound simple but it is all about real world experiences, research techniques and being able to critically analyse these compared to what they see in the AI world. Let's explore three areas in more detail;General Knowledge, Research Skills, and How thing work.

KID FRIENDLY
VERIFICATION CHECKLIST
S - E - F - T - I (SAFETY)

SOURCE
Who made it?
Are they Qualified?
Can we find and "About" Page?

EVIDENCE
Do they show data, photos or a primary source we can check?

FRESHNESS
Who made it?
Are they Qualified?
Can we find and "About" Page?

TRACEABILITY
Can we follow links back to where the claim came from?

INTENT
Is it to inform, sell, shock or manipulate?

EXTRA TOOLS TO TEACH

- SEARCH THE EXACT HEADLINE IN QUOTES - DO REPUTABLE OUTLETS REPORT IT?
- REVERSE IMAGE SEARCH - DOES THE IMAGE APPEAR YEARS EARLIER OR IN ANOTHER CONTEXT?
- LATERAL READING - OPEN A NEW TAB AND READ ABOUT THE SOURCE. BEFORE BELIEVING THE SOURCE.

FAMILY DATA BASICS

THESE DATA BASICS HELP KEEP YOUR CHILDREN SAFE ONLINE.
MATCH THEM WITH YOUR FAMILY DATA-SAFETY PLAYBOOK AND YOU'VE COVERED
MOST ISSUES, BUT STILL BE VIGILANT WHEN CHILDREN ARE ONLINE

PRIVACY FIRST
Dont share your full name, birthday, school, home address or live location publicly

STRONG LOGINS
Unique passwords and when available use two factor authentication.

BACKUPS
Keep school work in two places (eg cloud and USB)

ASK BEFORE POSTING
If a photo includes others, get consent

GRANDMA RULE
Would you be happy for your Grandparent to see the photo or comment.

PAUSE AND CHECK
If a message sounds urgent, scary or secret; stop and verify with a trusted adult.

FAMILY DATA-SAFETY PLAYBOOK
KID SIZED RULES

1. Never use technology privately, always use technology in the living room, lounge room or common areas.

2. Ask before you post anyone's face or voice.

3. No sharing full name, birthday, school, home address, or live location publicly.

4. Use strong passwords; don't reuse them; turn on 2FA where possible.

5. Back up schoolwork in two places.

6. Trust your feelings: If a message feels pushy, secret, or scary—stop and check with an adult.

7. Code word rule: Families can agree on a private code word for phone calls to avoid voice-clone scams.

8. Screenshot & save: If something seems off, keep a record; then verify.

💬 GENERAL KNOWLEDGE

General knowledge acts like a **BS radar.** Children won't remember everything, but they can use the shape of the world in their heads so they can test whether AI's answer *fits.* For all ages the easiest way to teach general knowledge is to talk with your children and experience the world. Take every opportunity to explain things to them whether you are having a discussion at the dinner table, or out walking in the park. Use these opportunities to explain how life works and how science fits into the world. If they ask about how rainbows are formed, then you explain in detail using science terminology and concepts. If you aren't sure of the details yourself, use your own research skills and use the tools you have like google to find out. Going through the research process also gives you the opportunity to show them how to find out interesting information. You can then show them how to analyse the information to see if it makes sense from what you know.

💡Parenting Tip: I often found myself and the children going out for the day to fill in our time. These were unique opportunities to experience the world and learn more about it. We would go to a museum, art gallery or a zoo. We even use to go on drives in the afternoon to find different animals around our area. I knew where all the sheep, cows, dogs, alpacas and goats were within a ½ hour drive. On her 2nd birthday I asked her what animal she wanted to see. She happily exclaimed "A giraffe". I was flabbergasted, however since it was her birthday we went to the zoo. We ended up with a yearly pass which we used most weeks to go and see 1 or 2 animals. If you have a zoo near you, I highly recommend it. The hippopotamus enclosure was her favourite and she even got to hold the hose to wash the hippo's teeth one day because the carer recognised us. Today you would have to pay a lot of money to get that experience.

When you are at these places take the time to read the plaques and after when you are relaxing or having food ask your children questions about what they learned. The first time, they might not know much, but the next time they won't be caught out by Mum or Dad and they will try and remember more. Maybe they could ask you questions about a certain exhibit.

Don't worry if they don't remember every single detail they learned in the day. The more you go, the more they will retain. It definitely gives them a good general knowledge and understanding so that when the detail is forgotten, they still know the basics and can work out if the AI answer is telling the truth or fake.

Children are wired to make you happy

Yes you heard me correct. Children's brains are wired to want to please you the parent. It's a developmental instinct which initially helps them not starve. It is instinctual and all children have it. I'm sure you are saying " yeah, my kid just wants to do the opposite of what I want".This may be true for older children but they generally are doing these naughty things to get your attention.

Let me explain a little more. When children are 6 weeks and they smile, you smile back and laugh. It is at this point that they realise that they can get your attention when they smile and giggle and not just cry.They like this, so they do it more. As they get older, they want to make you happy. Making you happy means that they get fed and clean. In their instinctual way they do more things which make you smile or laugh. I'm sure you have inadvertently laughed at a rude word your little one has said and then they just say it all the time. The only way to stop this is to not laugh (which is impossible sometimes).

The best part of this instinctual behaviour is that, if you pay attention to something they do, then they will do it more. We can use this to our advantage. If you want them to be polite, then pay attention to manners and notice when they are polite. If you want them to be good at sport, then pay attention to their abilities in sport and watch sport together. That way they know that sport is important to you and they will strive to do their best. If you want them to be academic, then have academic discussions at the dinner table and notice when they remember facts or work out problems.

If you want them to be good at all of these at the same time you can. There are plenty of hours in the day which you can show your children that you are interested in a topic. It's not limited to one topic. It's all about where **your** attention is. Is it on your phone when they are playing their sport game or playing their instrument? Is it on what they are doing? If it's on your phone, then they won't try as them playing the game or instrument doesn't make you happy.

Take advantage of biology and instinct. Know that they want to make you happy and help them to be the best person they can be.

General Knowledge through Travel

Travelling around your city, state or country can open children minds to how our cities work, how other people live, what other people have in comparison to them and how different people from different cultures behave. Traveling to other countries gives children a hands on understanding of how different cultures live. At this age they can recognise the difference between cultures and how people live. It can have a wonderful impact on their emotional intelligence as well as their academic intelligence.If they don't remember all the details, children remember the feelings, smells, voices, noises and the weather they had when they travelled. Remember to take lots of photos as they will re-live the specific moments you have captured through pictures and stories. They will also remember their own details from the trips as well as the collective memories, photos and stories of the family. Talking about the trip after is key to remembering the details and adding to their general knowledge.

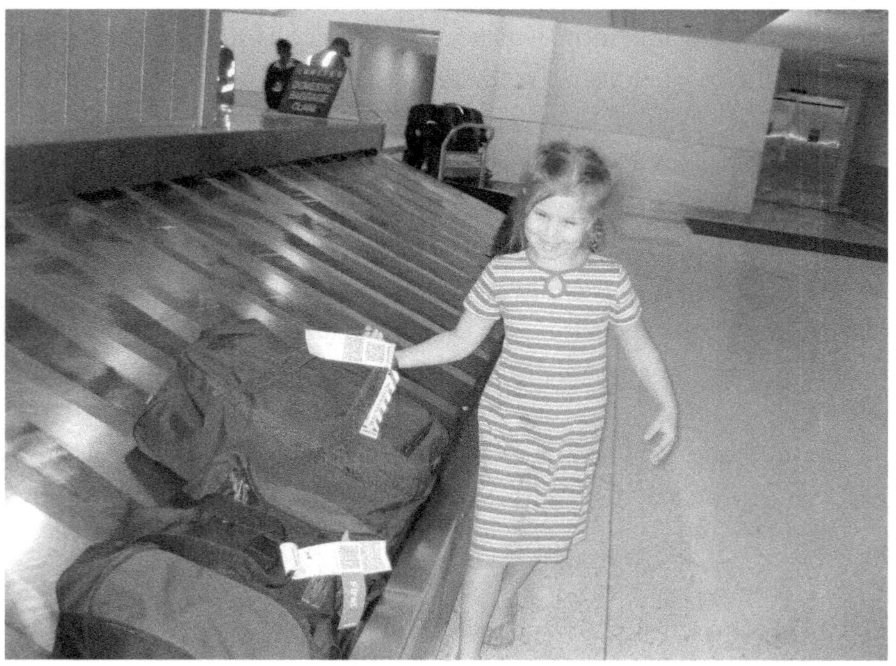

💡Parenting Tip: Whenever we travelled overseas I always made up a photo book with anecdotes and pictures. As adults, they still love coming over to my place and looking through the picture books.

Travelling overseas when they are young will give them a life filled with wonderful memories of the ways different communities prioritise different values in all sorts of countries. It peaks their curiosity and gives them an understanding that the world is really a big place, but we are all human. All of these experiences give a plethora of data so that they can use and analyse it when dealing with AI in their future. It helps them to distinguish fact from fiction. Analysing the information you are given from an AI is of utmost importance and when travelling you start to understand more about how communities and countries work differently.

Memory

General knowledge is built through remembering facts and reliant on our memory. It is very rare to have an ability to memorise without repetition, like those with a photographic memory. The rest of us need constant repetition. If you are not accessing or using information regularly our brain may just get rid of it. Remember when I mentioned neuroplasticity? Those pathways we create are often leading to memories and the less we walk those paths the harder it is to remember. How many times do you say to yourself "I knew this at high school" about a maths formula or history fact. Sometimes it feels like any fact which you haven't used since high school is gone into the abyss of your mind never to be retrieved. For this reason it is very important that children are constantly reminded of information they know.

There are many ways to keep including repetition of general knowledge with your children. My personal favourites are playing children trivia games, watching documentaries together, and talking about children news such as BTN (Behind the News on the ABC). All of these fun things you do together can build their general knowledge. Even just talking about different topics at the dinner table can be a quick way of learning something new. Finding out about the interests their friends have and doing a research project on it so you can talk about it with them and empathise with them is powerful. Giving them a **love of learning** in life is one of the most precious gifts you can give your children in our quickly changing world. If you value this they also will want to have a good general knowledge. You can show them that it is important by being interested in any new topics they have learnt.

Why memory fades

Not all of us have brains which can remember everything that happens. Brains prune unused facts. Short, frequent refreshers beat cramming every time. Constant repetition and revisiting facts will help memory stick. Your interest is the secret glue—when you care, kids try to remember.

Fun activities to support General Knowledge

💬 **Talk and notice:** At dinner or on a walk, explain how things work—weather, traffic lights, voting, banks, sports rules. Use real terms and plain examples.

- *Skills Built:* Observation, curiosity, reasoning, and applied understanding.
- *AI-World Connection:* When parents explain everyday systems, from traffic lights to voting, they're building children's critical thinking and general knowledge. This habit helps kids see how systems connect, just like AI learns from patterns. These "micro-lessons" develop the analytical thinking and contextual awareness needed in data science, civic technology, environmental planning, and systems design, fields where curiosity and comprehension fuel innovation.

🏛 **Museums, galleries, zoos:** Read the plaques, then ask one or two questions later ("What surprised you?" "What would you change?").

- *Skills Built:* Questioning, interpretation, reflection, and cultural literacy.
- *AI-World Connection:* Reading plaques, asking *"What surprised you?"*, and discussing later transforms passive visits into active learning. It strengthens memory and curiosity which is the foundation of lifelong

learning. These reflective habits translate to future careers in AI curation, museum technology, research, and creative storytelling, where data meets discovery and human interpretation gives information meaning.

🏭 **Micro-field trips:** Explore your city, nearby towns, factories' visitor centres, farms, water treatment displays, or power stations' info hubs.

- *Skills Built:* Curiosity, environmental awareness, critical thinking, and application of knowledge.
- *AI-World Connection:* Visiting farms, water treatment plants, or power stations gives children a systems-level understanding of how the world works which is a crucial perspective in the AI age. It mirrors how engineers and analysts study real-world data flows. These explorations build interest in sustainability, urban design, clean energy, engineering, and AI-driven infrastructure management, where observation and understanding create smarter solutions.

✈ **Travel with reflection:** From ~10 years old, kids remember more and compare cultures. I make photo books with captions and little stories after every holiday, that way we can revisit it together so memories "stick." They love looking through these books even now as adults.

- *Skills Built:* Global awareness, empathy, perspective-taking, and storytelling.
- *AI-World Connection:* Travel photo books and reflection journals help children make sense of new experiences and connect across cultures which is a critical soft skill in the global AI economy. They learn how to document, compare, and communicate meaningfully, just as professionals in international development, global education, intercultural communication, and digital media storytelling do. AI may

translate language, but humans create connection.

↪ **Repetition matters:** Trivia games, documentaries, kids' news (e.g., BTN), and 5-minute dinner-table "teach-backs" keep facts fresh.

- *Skills Built:* Memory, comprehension, curiosity, and confidence.
- *AI-World Connection:* Trivia, kids' news, and dinner-table "teach-backs" turn repetition into mastery. Just as AI models strengthen through repeated learning cycles, children reinforce their neural pathways through recall and review. This builds the cognitive agility needed for research analysis, journalism, science communication, and education technology, where knowledge must stay sharp and current.

💭 **"Explain it like I'm 6/12/16" rounds:** One family member teaches a tiny fact at three levels.

- *Skills Built:* Simplification, communication, empathy, and adaptive thinking.
- *AI-World Connection:* Teaching a single fact at three age levels develops the ability to reframe and adapt messages which is a vital skill in AI education, leadership communication, UX design, and public engagement. The clearer children can explain an idea, the better they'll be at collaborating with humans and machines that depend on clarity to learn.

🏙 **City systems bingo:** Spot and label systems (storm drains, substation, bus depot).

- *Skills Built:* Observation, pattern recognition, and systems thinking.
- *AI-World Connection:* Spotting and labelling systems like bus depots or

storm drains helps children see how infrastructures interconnect which incidentally is exactly how AI models learn relationships between data points. This curiosity about how cities function builds foundations for careers in urban analytics, logistics, AI modelling, and smart city design, where every system tells a story.

Caption club: Pick one travel photo and write two true captions and one false—others guess the fake.

- *Skills Built:* Creativity, truth verification, humour, and critical thinking.
- *AI-World Connection:* Writing two true and one false caption teaches children how to spot misinformation and engage others critically. It's a playful introduction to media literacy, fact-checking, and digital storytelling, essential in future jobs in AI communication, digital ethics, journalism, and information verification.

BTN: Watching Behind the News on the ABC and talking about the news items on there is a great way to help build General Knowledge of children.

- *Skills Built:* Media literacy, global awareness, discussion, and comprehension.
- *AI-World Connection:* Watching BTN and discussing the news builds informed, curious thinkers who understand world systems. Children learn to distinguish fact from opinion, the same process AI tools use when classifying data. This builds readiness for roles in data ethics, education, public policy, and responsible media technology, where informed citizens shape responsible innovation.

Friend Research Project: Find out about the interests of their friends and do a research project on it. This way they can talk with their friends about their interest and empathise with them. This is a powerful gift to teach your child on how to build relationships.

- ***Skills Built:*** Empathy, curiosity, communication, and social intelligence.
- ***AI-World Connection:*** Researching a friend's interests develops not just knowledge but connection which is an essential balance between information and humanity. It models the empathy and interpersonal insight needed in AI ethics, psychology, user-experience research, and community leadership, where understanding people is just as important as understanding data.

AGE	ACTIVITY	EXAMPLE
1 – 5 years	Talking to your child about the world around them	
	Board Games	First Orchard Games, Zingo, Pete the cat and the missing cupcakes, Candy land, Count your chickens, TIPS: 10-15 minute games celebrating participation rather than winning
5 – 10 years	Board Games	Trivial pursuit children games Brain box Outfoxed Ticket to Ride: First Journey Timeline: Inventions or Events Wildcraft! A Herbal Adventure Game Professor Noggin's Card Game Brain Games - children version National Geographic children version Qwirkle Scrabble Family Feud
10 + years	Board Games	Wit's End, Timeline Challenge, Ticket to Ride, Catan, Pandemic, Brainiac Game Show, Terra, Smart Ass, Celebrity Heads, Family Feud, Game of Life, Trivial Pursuit Homemade board games - Children create their own board game based on a topic.

⚖ HOW THINGS WORK

The point of difference between human and AI is the skill of making connections between seemingly unrelated subjects. This particular skill has created great art works, feats of engineering greatness and the creation of AI. But we as a society are not finished yet. Now we can use AI to gather information to use in our new explorations. We can use AI to solve some of our more pressing problems we have as a society, but we cannot use it to change us as humans; our emotions, our creative thinking, our innovation, our desire to help others. GenAI, as they are calling them, need to know what makes sense, what is rubbish, what is a truth and what can't possibly physically happen when they are working with AI. Teaching our children how to research early on with an inquiring mind leads to innovation and creativity. These skills are essential as our lives change so quickly with the new emerging technologies.

Curiosity is a trait which children have in abundance. I hope that they never lose it. It is essential that they keep that curiosity in a world where AI can do most things for us. There are robots for cleaning, programs for shopping and AI guidance for whatever we put our minds to. Keeping that curiosity of Why, How, When, Who and What will help them to work collaboratively with AI to succeed in the new AI world.

When teaching children I love watching them learning about a new topic. You might have heard of the expression "children brains are like a sponge", this is so true especially when they are learning about new topics. At school we started to explore how packaged food is processed. We watched how chips went from potatoes to a packet of chips. Giving them that small bit of information sparked the curiosity and they were off learning all about the new subject. They couldn't get enough. Questions streamed from their brains. If chips are packaged like that, how are Pringles packaged? What about tins of tuna? Came another question. I couldn't keep up. This is exactly what they need to live in the future AI world. Their search for new knowledge will help develop the new pathway that humanity is taking. By teaching them all about the world around them; from how mechanical things work to what the process of design is will set them up for success in an AI world.

It's as easy as living life

Sometimes, it's just about talking or showing children how things work in their own world. They are curious about what you do at work, so talking to them about your work, or even going on a day trip can be fun. They don't need to understand it, they are just curious, and the more you feed that curiosity, the more questions flow. When you are on holidays you can go to the local factory. This will open up a whole new avenue of curiosity, questions and exploration. Even using a flow chart to explain how things work, can be very helpful for them in the future. The more experiences they are exposed to, the more curious they have and the more knowledge they gain, enabling them to analyse the truth of AI solutions. Kids who **know the world**, **ask good questions**, and **verify** before believing. They will steer AI rather than being steered by it.

If a house is being built near your house, then go everyday for a walk and watch all the parts being put together. It's fascinating and when they read books or see video's they will have a real life experience of how it is put together. I love Richard Scarry books to show children how towns and life in a town works. They give detailed pictures on how things work in a fun and cartoon way. The characters have adventures in BusyTown, learning how things work and helping to fix the issues that arise. They are problem solvers and think critically. This is the best introduction to this type of thinking and knowledge. The more fun it can be, the quicker they learn.

Do your own research and find similar books, maybe tailored to your children interests. Maybe they like cars and trucks or want to learn more about ecosystems, there are plenty of books out there which are both fun and educational.

DIY along side you

Let children help you DIY. Even if it is just getting them to pass tools or getting you a drink. It's important to remember that safety is the first priority and they must learn this early. All of these experiences help them later in life and it is a great help for you as you are DIYing.

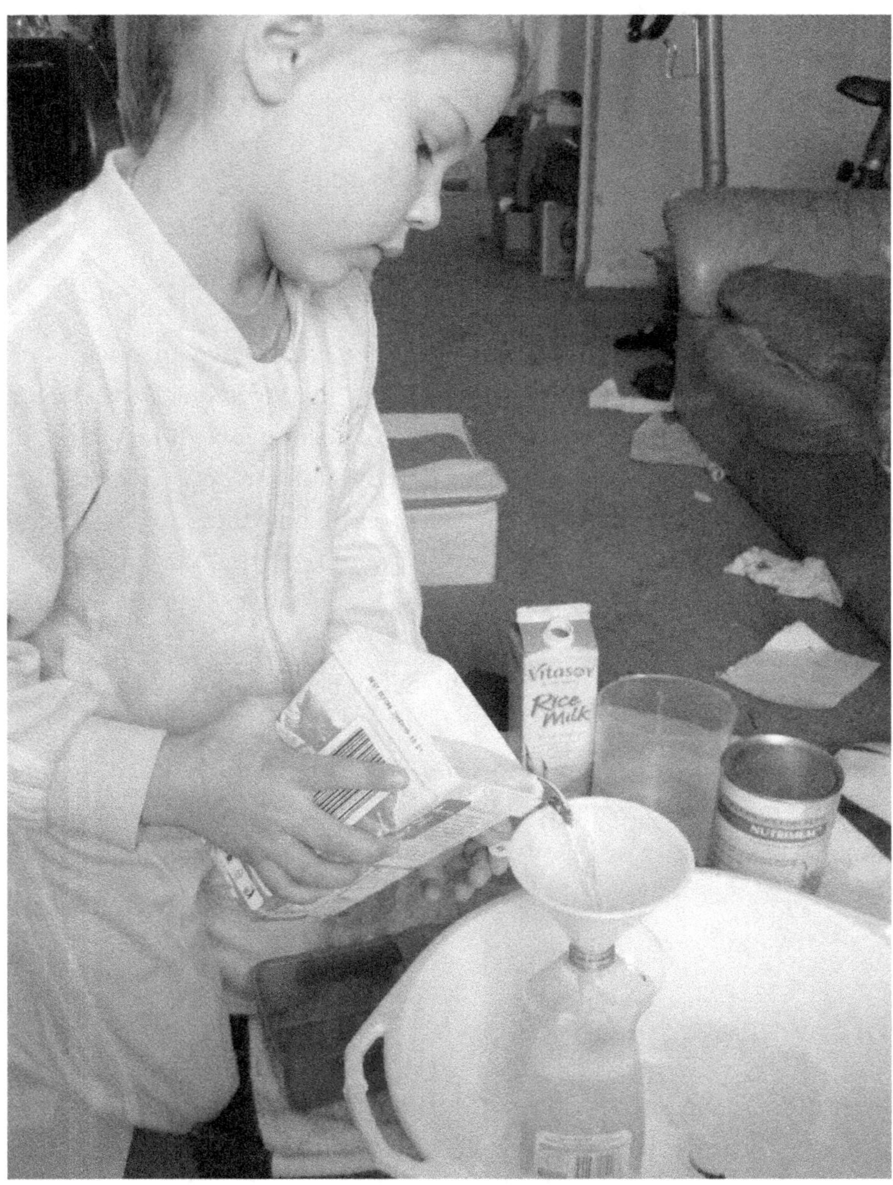

Children can even just watch what the plumber does so they have an idea on the processes which they use to fix the plumbing. It doesn't have to be you all the time. When they get older get them to help problem solve the situation with you. My daughter had a plumbing issue at her house recently where one of the boys had driven into the tap outside and knocked it off. The boys were hopeless as they hadn't had any experience in taps or plumbing. My daughter ran to the front of the house and turned off the water. Went on a trip to Bunnings to get the parts needed and fixed it for them. All those hours watching and helping me around the house had paid off.

We bought a new house while my children were teenagers, it came with no cupboards in the bedrooms. I decided that this was a great life lesson, so I got them involved. I could have just got someone in to design and make the wardrobes but I knew it would be a great lesson for their future. This was an opportunity I couldn't miss so the decision was made that the children could design their rooms including the wardrobes. They were already somewhat prepared, knowing how to research and budget from when I previously gave them the opportunity to help furnish one of our previous rentals. I must admit I had lots of input into the process by asking questions, pointing out problems which might happen and all of us watching lots of YouTube clips and renovation shows. They finished the design and we went on a trip to IKEA. Myself and the children build over 20 cupboards over a couple of weeks. My youngest daughter and I got so good at making them up we can still work seamlessly to put anything IKEA together. Our next project was successfully installing new kitchenette cupboards. The take away from this was how much of a hands on learning experience they got. Their understanding of building and how much precision is needed to make a product successful was priceless.

They can now use this experience to guide any prompts they make in AI with regards to building and house maintenance. The best thing was the fun we had together while problem solving issues. We laughed our way through missing screws, having to undo and re-do parts, the importance of reading the detail and so much more. Laughter definitely got us through. I kept saying "No point getting frustrated or angry, let's just fix it.", while laughing at how silly the mistake was. Believe me there was also lots of chocolate.

Kitchen engineering

Even as little ones, my children would sit on the kitchen bench and watch me cook. As adults they have a real love for cooking, trying new recipes and old ones my Nanna had handed down. The repetition and camaraderie during the process of cooking made it enjoyable for everyone and helped them to expand their palette. It was a real learning experience for them and gave me company as I spent many hours in the kitchen cooking meals. In our household there are lots of allergies which has meant that I had to cook from scratch most days. It was the perfect repetition and real-world feedback they needed for future success. This was the silver lining to limited fast food.

Roadside reality

Flat tyre? Talk through the manual, the jack point, and the "loosen on the ground first" trick. Involve them by getting them to hold things when you are changing your tyre. Give them the tips to make it work easier, like 'keep the tyre on the ground and loosen the screws before lifting it off the ground'. These tips and tricks are things they will remember. Involve them by getting them to hold things. They might complain but the more you get them to help the less resistance you will get in the future. Later, that knowledge drives safe, calm problem-solving. You'll be glad when it's 10 o'clock at night and they rock up home safely saying "We had a flat tonight. The spare is on and everyone is home safe. Oh, and we need to fix that tyre tomorrow Mum."

Inventiveness in action

My 5 foot aerialist daughter solved the "too-tall rig" problem using the car's tow bar and ropes to tip and slot sections—a practical masterclass in leverage, sequencing, and safety. It's genius, getting the car to do the heavy lifting and it allows her to put her rig up anywhere, on her own, with very little help. I'm sure all the times we used the tow bar on the car to pull trailers and pull trees on the farm helped her to problem solve this issue herself. It shows how worldly life experiences can't be under estimated or replaced. Life experience breeds elegant solutions.

These are just a few examples of helping children to understand how to life works and how to fix things. It helps their critical thinking skills and problem solving skills. The results of our hard work teaching them these skills can benefit their inventiveness. My 5 foot daughter is now an aerialist performer and teacher. She recently came across her "too-tall rig" problem. The equipment is 7m split up into 1.2m length heavy sections and is transported only on the roof of her car.. She solved her problem with a practical masterclass in leverage, sequencing, and safety. Her approach was to attach rope to the tow bar on the car to tip the top of the rig slightly to slot the new leg pieces in place. Once it's safe she then release the tension on the rope,by reversing the car, so it can be tipped the other way to put the next 2 legs on. Driving back and forth tipping the rig to and fro until all legs were are in place. It's genius, getting the car to do the heavy lifting and it allows her to put it up anywhere, on her own, with very little help. I'm sure all the times we used the tow bar on the car to pull trailers and pull trees on the farm helped her to problem solve this issue herself.It shows how worldly life experiences can't be under estimated or replaced. Life experiences breed elegant solutions.

Why this all matters in the AI world

Take every opportunity to teach your children how the world works around them. You won't regret it. Having an inquiring mind will help them to ask good questions and verify AI data. AI needs a human brain to wonder and understanding what is involved with solving a problem.This curiosity and enquiry is what will set us apart from AI.

Fun activities to support How things work

⟍ Fix-it Fridays: Choose one tiny repair or maintenance task together.

- *Skills Built:* Practical problem-solving, responsibility, patience, and curiosity.
- *AI-World Connection:* Choosing one small household repair each week, like tightening a hinge or replacing batteries. It teaches children how systems fail and how to restore them. This builds an engineer's mindset: diagnosing issues, testing solutions, and valuing maintenance. These foundational habits connect directly to future careers in robotics, hardware engineering, AI maintenance, and sustainable technology, where hands-on understanding and iterative thinking are crucial.

Deconstruct day: Open an old toaster (unplugged), name parts, sketch how electricity flows.

- *Skills Built:* Curiosity, analytical thinking, observation, and design awareness.
- *AI-World Connection:* Taking apart an old appliance and identifying its components transforms children into mini-engineers. They learn that every system, from a toaster to a smartphone, is made up of interconnected parts, just like AI models. This activity lays the groundwork for mechanical engineering, product design, and AI hardware innovation, where curiosity about "how things work" drives technological progress.

Systems work: Map how water gets to your tap and where it goes after the sink.

- *Skills Built:* Systems thinking, environmental awareness, and cause-and-effect reasoning.

- *AI-World Connection:* Mapping where water comes from and where it goes after use helps children visualise unseen networks which is exactly how AI analyses invisible data systems. They begin to grasp interdependence and flow, essential for environmental science, infrastructure analytics, sustainability engineering, and AI-driven urban planning, where data and physical systems merge to create smarter, greener solutions.

Build & test: Paper bridges or straw towers; change one variable at a time and record outcomes.

- *Skills Built:* Experimentation, scientific thinking, patience, and data collection.
- *AI-World Connection:* Building paper bridges or straw towers and testing different variables introduces the principles of the scientific method. This methodology is also used by AI to test, refine, and learn from data. It teaches children to hypothesise, measure, and iterate. These habits translate into success in research, structural engineering, AI modelling, and innovation design, where experimentation and iteration are everything.

⚙ **Cookbook lab:** Double/halve recipes; predict and observe differences.

- *Skills Built:* Measurement, prediction, observation, and cause–effect reasoning.
- *AI-World Connection:* Cooking is a living lab for logic and pattern recognition. Doubling or halving recipes builds mathematical reasoning, while predicting outcomes strengthens analytical thinking. This "tasty science" forms the cognitive base for future work in data analytics, chemistry, AI simulation, and algorithmic design, where small changes in input lead to big changes in outcome, just like in the kitchen.

AGE	ACTIVITY	EXAMPLE
1 – 5 years	Watching Construction	Observation of houses being built, or apartment blocks
	Pretend play Toys	Build a Fort, Play with a dolls house, Make a cubby, Pretend tool kits, Toy kitchens, Toy cleaning
	Construction toys	Duplo, Meccano, Magnetic Squares, Mobilo, Wooden blocks, k'nex etc
	Richard Scarry books	
	Cooking together	
	Chores	
	Visit museum and art galleries	
	Visit zoo	
5 – 10 years	Construction toys	Duplo, Meccano, Magnetic squares, Mobilo, wooden blocks, k'nex
	DYI helper	They can watch you and hold tools, get water etc
	Go on factory tours	
	Holiday at a farm	
	Try out a steam train and learn all about how they work	
	Board Games	Monopoly – learn about money, buying and selling Risk - global conflict
10 + years	All prior activities are relevant	
	DYI - building flat packed furniture	
	House hold maintenance	It's about knowing what needs to be done and who to call.
	Car maintenance	Change a tyre, fill with water, fill with oil
	Holiday overseas	

✎ RESEARCH SKILLS: FROM CURIOSITY TO CREDIBLE ANSWERS

We have been talking about the difference between AI and humans so that we can understand what skills our children will need in the future to be able to take advantage of the new technology and the new jobs which are being created. I believe that one of the most important points of difference between us and AI in the job market is the skill of making connections between seemingly unrelated subjects. This particular skill has created great art works, incredible feats of engineering greatness and the creation of AI in the past. But we as a society are not finished yet. Now we can use AI to gather information to use in our new explorations. We can use AI to solve some of our more pressing problems we have as a society, but we cannot use it to change us as humans; our emotions, our creative thinking, our innovation, and our desire to help others. The new generation need to know what makes sense (truth), what is rubbish (fake) and what can't possibly physically happen when working with AI. Teaching our children how to research early on with an inquiring mind will support them in this endeavour as well as leading to new innovation and creativity. These are the skills which are essential for our children to master as our lives change so quickly with the new emerging technologies.

Right now, AI can help with **research**, but it doesn't replace having a question or judging the result. We can teach children to move from wondering to checking then deciding. Currently we can use AI as a research tool by asking it to do the tasks needed in a research project, however it cannot yet come up with a research topic on it's own. It may be closer than you think, but this is one of the big differences between AI and human. Coming up with a research topic requires creativity, critical thinking skills and innovation which we can definitely teach to children.

Boredom is a state of mind

It is said that **creativity** is built through boredom. Yes. That's right. You need to let your children to be bored so that they can learn to be creative and develop new innovative ideas. Don't get me wrong, it's not just about leaving them to be bored, they also need to have the skills to accomplish the idea, but the creative innovative ideas all come from a place of boredom and silence. Have you ever been in the shower or woken up with a brilliant idea? It's about letting the mind wander and come up with connections between unrelated topics.These ideas can be a breakthrough in engineering. Having the problem mull around in your mind helps you to come up with that brilliant solution. Wondering about how a bridge can hold so many cars can lead to a new bridge design. It starts off small, with small problems and then grows.

I have seen engineering designs which help collect trash from the ocean. All designed by a young man. He saw the problem, wondered how he could solve the problem, researched the solution and has developed this new boat solution which can collect floating rubbish from the oceans of the world. All from being bored.

When children complain to you about being bored say "Boredom is a state of mind. Change it and find something to do". My kids quote this all the time. They were sick of me saying it, but they took it on and started to create incredible inventions.

One holiday we were at the family farm. When we were staying there, I normally gave them a lunch pack and told them to come back before the sun goes down as they knew how to stay safe there. This particular day they started saying "But there is nothing to do here. It's boring". "Boredom is a state of mind, so find something to do." was my response. They walked away in a huff, however they came back 1 hour later and asked for scissors, string and a small saw. I was intrigued but I left them to it. When I hadn't heard from them in another 2 hours, I went for a walk to find them. They had created their own Hunger Games arena in one of the paddocks. It was amazing. They had created their own bow and arrow (hence the string) and they played there for the rest of the week trying to catch hares and birds for lunch. This creativity was priceless. This is the perfect example of how **boredom builds creativity.** So remember that it's ok for children to be bored as it gives their brains time to think and wonder, explore and be creative.

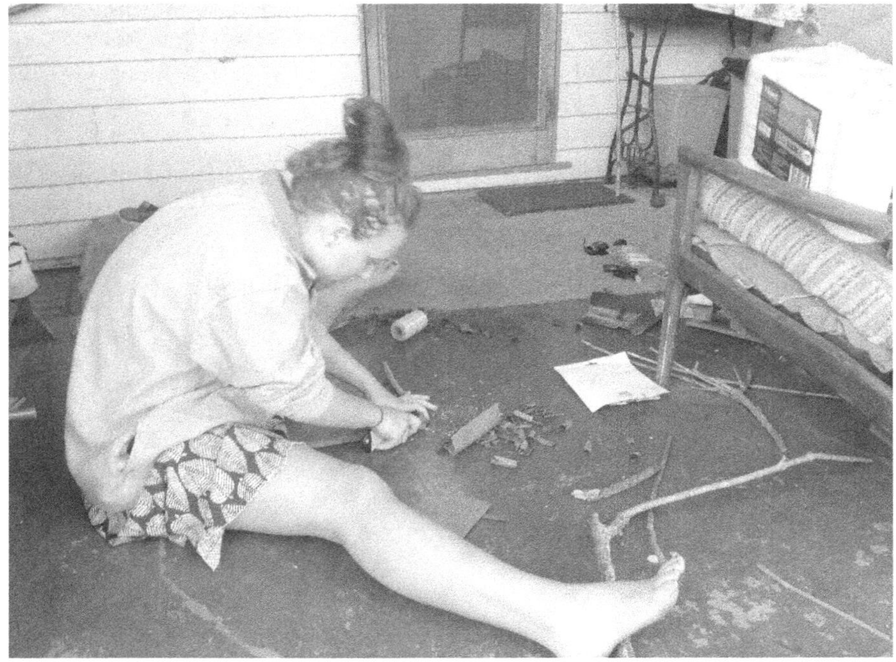

Fun ways to support creativity.

AGE	ACTIVITY	EXAMPLE
1 – 5 years	Sand pit play	Sand pit toys, buckets, spades, pots and pans
	Open-ended exploring play	Just allow them to play with anything in the house or garden as long as it's safe. Their imagination will take over.
	Experimenting and discovering how things work	
	Imaginative play	with dress-up clothes and toys (kitchen set, laundry set, tool kit etc)
	Consruction toys	Duplo, Meccano, Magnetic squares, Mobilo, Wooden blocks, k'nex
	Crafting with recyclables	Beads, Paper, Recycling construction
	Puzzles	Up to 300 pieces
5 – 10 years	Finding things you have lost	
	Board Games	Pictionary, Cranium, Codenames, Telestrations, Dixit
	Craft	Origami, Beading, Woodwork or paper craft
	Sewing	
	Knitting	
10 + years	As per 5-10 years	
	Craft	Pottery
	Solving Mysteries	
	Reading	Mystery books, how to books
	Exploring new places	
	Learning how to take public transport	
	Experimenting and discovering how things work	Myth busters show is great for science and engineering based info

Create your own Research Projects

💡Parenting Tip: In our house we loved researching topics and making them into projects.

Before we went on a trip to Ballarat we researched what we could do there. I did this every big holiday as it builds excitement for the holiday and helps them to be engaged when there. They found out about Sovereign Hill, an open-air gold mine museum which presents the story of Ballarat as a gold rush boom town. We researched the time period it was set in and found out about the gold rush, candle-making, shops, and soldiers. When we got there we saw actors dressed up in tradition clothing of the time and everything they had learned about. They learned to make their own candles and got to pan for gold. One of my daughters even got to be a solider. It created a real life experience on top of the researched learning which meant that they has so much more fun. The visit clicked because the research had primed their minds. They made connections between "then" and "now" and remembered much more. It was a magical experience for them because of their research. I have to say though, that their favourite shop was the candy shop.

Here are the girls hanging out with soldiers and panning for gold

Teach with real errands - Furniture project

Research is all about analysing the problem, coming up with a few ideas, then being able to think up the solution for the most relevant one. This can be done together. Get your children involved in basic research at home. Where can I get a new dining table from? Give them ideas on how to go about researching and finding the answer.

When we lost all of our possessions to a big storm, I did not want to go through finding all new furniture again. I decided it would be a good lesson for my teenagers to learn about budgeting and research. I gave them a budget and the basic types of furniture we needed and let them go. They researched all different types of furniture they could buy, the price, the size and if it came as a flat pack. After their research they presented me with their list of places to go looking for furniture and we all trudged around trying to find something we liked. After an exhaustive lot of research, they finalised on the items and I bought them.It was a great way that they were able to learn how to research and find out where to buy furniture. They tell me that they use these skills all the time now as adult's, that they appreciate the opportunities I gave them as teenagers.

Support an Inquiring mind

Research is all about having an inquiring mind which understands how the world works. Helping our children to learn about the working of the world around them will give them the tools they need to make the most out of the new AI technology. It is all about thinking and asking questions. The tedious data analysis and gathering for research can be done by AI if you know the prompts to use. If your children have had the opportunity to become creative, innovative, problem solvers, during their childhood, it will lead to job offers in the future. Being able to use AI and think outside the box will be the difference between being successful in the future and just becoming a statistic.

Fun ways to support Research Skills

☝**Question Jar:** Pull a question at dinner; spend 5 minutes finding two sources; share what matched and what didn't.

- *Skills Built:* Curiosity, verification, comparative reasoning, and analytical thinking.
- *AI-World Connection:* Pulling a question from the jar at dinner and checking multiple sources teaches children that not all answers are created equal. This habit mirrors how responsible AI systems cross-check data before giving a response. It helps children develop data verification, critical research, and discernment, vital skills for future roles in AI training, journalism, education technology, and research analysis, where truth depends on comparing evidence and context, not just accepting the first answer.

⚖ **Source Showdown:** Each person brings one source on the same topic; vote on credibility using the checklist below.

- *Skills Built:* Critical evaluation, collaboration, argumentation, and ethical judgment.
- *AI-World Connection:* Having each person bring one source and then voting on credibility introduces children to evidence-based decision-making. It's a playful way to apply the S.E.F.T.-I. (Source, Evidence, Freshness, Traceability, Intent) framework from your earlier chapter. This helps them think like AI auditors who learn how to filter bias, spot misinformation, and prioritise quality data. These skills connect directly to careers in AI governance, data ethics, law, and investigative research, where knowing which information to trust defines success.

🎥 **Mini-documentary:** children script and record a 60-second video teaching one thing they learned, citing sources on-screen.

- ***Skills Built:*** Research, storytelling, synthesis, digital communication, and accountability.
- ***AI-World Connection:*** When children script, film, and cite sources in a one-minute documentary, they learn how to transform data into insight which is a critical skill in the AI era. This integrates creativity with evidence, just as professionals do when presenting findings visually or through media. It builds confidence in digital content creation, science communication, data storytelling, and educational media design, where AI tools assist with editing, but humans bring narrative and meaning.

AGE	ACTIVITY	EXAMPLE
1 – 5 years	Sorting toys	Sort by colour, size or type
	Mystery Bag	Child has to guess the toy in the bag
	Science experiments	Eg. Will it sink or float?
	Mini scientist	Draws and observes bugs or animals they can see
	Shadow puppets	With a torch shine on a toy and trace it
	Yes or no detective	You think up an object and the child has to ask questions to determine what it is. You can only answer yes or no.
	STEM projects	Microscope, engineering wheels, national geographic toys, lava lamp, making slime
	Weather watchers	Observe the weather, record and make predictions
5 – 10 years	Mystery scientist	Find a science question and look up credible research books and online resources.
	Sorting and classifying	Use toys, rocks, shells, lego and ask to sort and classify into groups and explain their reasoning
	Be a Mythbuster	Watch the Mythbusters show and then make up your own myth to bust. eg. What type of paper plane flys furtherest? Make a hypothesis, run a test and then record your results
	Treasure hunt	You can make clues around the house and the children need to follow them to find the treasure
	Data detectives	Children think up a question and ask friends and family.They then make a chart and talk about it
	CSI: Crime Scene	Make a crime scene such as a toy going missing. Set out clues (Footprints, notes, snacks crumbs) and the children need to find evidence and make logical deductions.
	Microscope investigation	Test different materials under a microscope
	Research together	Research questions which are relevant such as what pet is the best? Use technology together, libraries, and other resources to answer questions they have come up with (habitat, diet, predators, space needed, time required etc) **TIP:** Model curiosity (let the children see you question things), Encourage reflection (ask what did you discover after the game/activity), Promote team work and collaboration, Introduce basic research tools (magnifying glass, microscope, clipboards, rulers, stopwatches, glasses)

AGE	ACTIVITY	EXAMPLE
10 + years	The Great Debate Challenge	Get into teams, assign a topic, research and evaluate the information, formulate arguments and present your findings on powerpoint. Make sure you include a neutral judge
	Consumer Detective	Research different brands of the same product and compare price, quality, ingredients and marketing claims. Present the findings in a consumer report
	Myth vs Fact	Research credible sources and present results in a infographic, powerpoint, TikTok video, or comic strip
	Historical Detective	Give clues about a historical event. Teens use books and online research to work out what and where you are talking about. Findings are to presented in a case report
	Community Research	Teens find a local issues, collect data, create a visual presentation
	Fact-Checking Olympics or World Records	Find a viral headline and see if you can find out if it is true or not. Teach CRAAP (Currency, Relevance, Authority, Accuracy and Purpose)
	Shark Tank	Develop a product. Research audience, competitors and costs then pitch their product TIP: watch Gruen transfer or shark tank shows)
	Escape room style puzzles	

💡 PROBLEM SOLVING

There are so many aspects to problem solving. We are not just talking about a math problem "When will the train arrive if it is going from Bundaberg to Brisbane at 100 km/hr and it leaves at 9am" No, we are talking about life problem skills as well. Although it is important to be able to solve math problems, real life problems help children to repeat the process and refine it. They could problem solve 'How do we get to the football if we are not driving?' and talking through the process of 'How can we choose what we are wearing to book week?'. Children learn these skills through living life. Sometimes they need to work through the problem themselves and deal with the consequences (e.g. didn't follow your advice and bring a jumper so they get cold) They learn the process of identifying and evaluating different options.

The simple loop children can learn

There is a standard problem solving process you can teach them, no matter what the problem is and it follows this loop:

- Analyse the problem
- Brainstorm Possible Solutions
- Evaluate the Solutions
- Implement the Chosen Solution
- Evaluate and Make Changes
- Reflect on the process
- Back to Re-Analyse the Problem

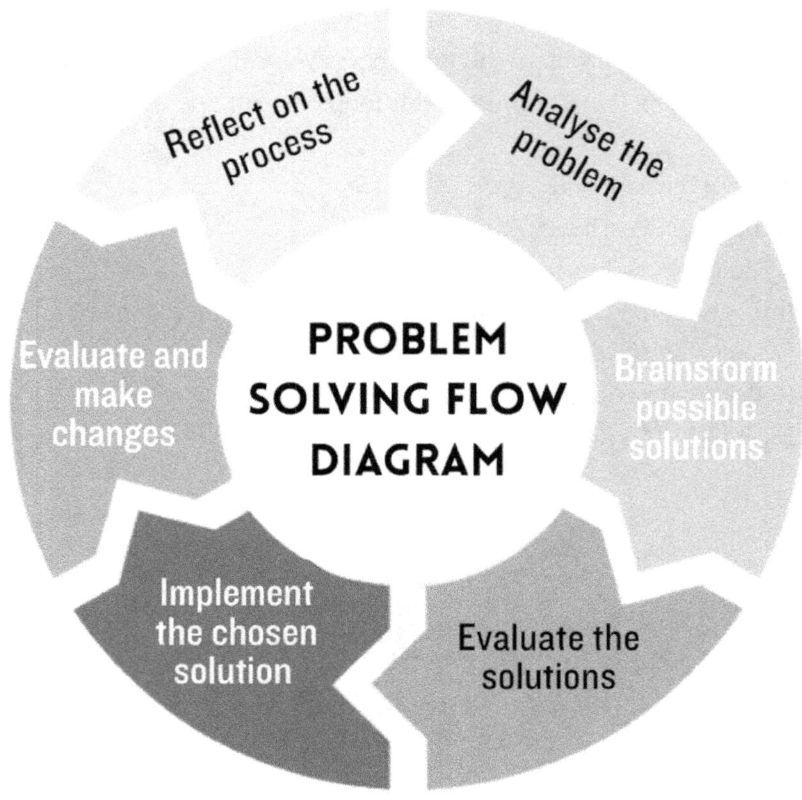

You probably run this loop automatically. The difference is that now you'll **say it out loud** so children can join in and eventually lead it themselves.

To effectively teach problem-solving skills to children, start by understanding this process. You probably do this without thinking, but we need to be able to talk it through to get our children involved. Sometimes just encouraging creative play, and using open-ended questions to guide their thinking can start the process of learning how to problem solve.

💡**Parenting Tip:** Remember that they need to **do** the process so standing back and letting them make the mistakes is often what is needed. Always be there to help out if all goes haywire and it gets dangerous, but children learn best by repetition and experiencing their own mistakes.

How to Model the Problem Solving Process

The best way to teach children how to problem solve it to do the process together as many times as you can. Once you are faced with a problem they can help with, break down the problem into small, manageable steps and encourage them to brainstorm lots of solutions with you. Then they need to work out which option is the best. I sometimes use the list of pro's and con's or even just put ticks beside the ideas which work best. Then you get to solve the problem by implementing the solution you have decided. Most decisions are based on the simple flow, so you should easily be able to explain it to the children. The more you do this, the easier it is for them to apply this knowledge to new situations. You will see them start to incorporate this process in their pretend play. One of the most important parts of this technique is giving them a chance to trial the process.

By providing opportunities for them to learn from mistakes and reflect on their own experiences they will learn the skill so much quicker. The developing brain likes to physically do things not just be told outright, but guided and then allowed to experience it.

💡**Parenting Tip:** Here are some tips on each step of the process.

- **Analyse the problem** - Use plain language: "What's the actual problem?". Break it into smaller bits if it feels big. Say "What's the real snag here—time, money, tools, or people?"

- **Brainstorm options** - Quantity over quality at first. "Let's list three ways, even silly ones." Ask open-ended questions: "What else could work?" "What would Bluey try?"
- **Evaluate Solutions** - Use quick tools: a pros/cons list, ticks next to workable ideas, or a 1–5 rating for effort/cost/fun.
- **Implement** (do it) - Say: *"Let's make this."* (Skip the jargon.)
- **Evaluate and Make Changes** - "What didn't work and can we fix it easily?" "Which one is fastest? Which costs least? Which is safest?"
- **Reflect on the Process** - "What should we keep or tweak next time?" What worked? What would we change? Celebrate the attempt, not just the outcome.
- **Loop back** - If it didn't work, that's information. Try the next best option.

Make Problem Solving Fun

We know that children learn best through fun, and play, so how do we do this in a fun way? Just involving them in the process helps.It is fun to do things with Mum and Dad. The only thing I will warn you about. If it matters to you the outcome don't re-do it or change it 'just because'. That will undermine all that you have done. Choose problems where you don't care about the outcome (or you can fix when they are asleep). They are clever little humans and understand when it's not worth trying or contributing because Mum and Dad just don't listen and do it their way no matter what. It's all about choosing the right problems and having fun with it as well as enjoying the outcome.

Keep it at their level

Remember that they are little and don't need to know the terms "implementing the solution" until later. Use simple terms like "Let's make this". As they grow they will learn the formal names for each step at school. By all means use the lingo but make it fun.

> 🔅**Parenting Tip:** Count to 4 before you give an answer. Sometimes their brains just need a little time to process and they come up with the answer, or even a better solution.

Stories - The Secret Superpower for Problem Solving.

Books are an essential learning material for skills like general knowledge & learning how things work. There are many books available which show lots of problem solving techniques in the stories too. We find that most stories have a problem which needs to be fixed. Bluey get's into lots of situations which need fixing and the Richard Scarry characters we have mentioned before are always fixing problems in the city. There are many children's cartoons which get into all sorts of mischief. The Disney channel is full of them. Choose the ones you like the best and let the children go through the problem solving process with the characters. It's so cute when you see them mimic the scenes in their own pretend play. This is their way of running scenarios to be confident in how the world works around them.

> 🔅**Parenting Tip:** Toys such as pretend tools, kitchens, wheelbarrows and sand pits are great to have around to enhance their play while they learn to problem solve.

As they get older their age appropriate stories and books will start to incorporate more details where characters go on adventures and solve problems. These are one of the best ways to teach children a different set of problem solving skills, which comes with their developmental stage such as friendship issues, or life problems. There are so many series of books which are based on topics which might interest your child. There are animal ones, girl adventures, boy adventures, funny stories, unicorn and fairy books to name a few.The list is endless. Amazon have a great sorting option where you can pop in the age of the child and they recommend books and series to read.

> 💡**Parenting Tip:** Don't discount the older books (such as Anne of Green Gables by Lucy Maud Montgomery, Famous Five by Enid Blyton etc) as their grammar and sentence structure is sometimes better than modern books and the children get the added bonus of having these skills accurately modelled.

As a mother of girls and a teacher of many years, I understand the trials of friendship problems. Having girls read friendship books help them to go through issues with friends in their books and gives them guidelines on what to do if they find themselves in a similar situation. They get to live the issues without actually going through the situation or feeling the emotions as strong. Don't get me wrong, they will feel the emotions with the characters, the frustration and anger if the solution doesn't work at first and the elation when it does, but it's not quite at the same intensity when you are an observer and it's not actually happening to you. This all builds their emotional intelligence in a soft and easy way.

Building toys for problem solving

I also love challenging children with problems in lego or building toys. As long as they have created a number of different models with the instructions and understand how to put together shapes in lego, give them a picture and tell them to use their lego pieces to make that picture. You will be amazed at how long they will try and create it exactly. It's a great way to practise problem solving as they won't have actually purpose made parts to use in the picture creation. There are also many shows you can watch together which solves problems. I love the Lego Masters program where people get to create models and you get to watch the whole problem solving, building, and creation process.

Practical ideas on how to model problem-solving behaviour:

Children learn best by watching you. By incorporating these strategies when you are problem solving with your children, you can help them develop strong problem-solving skills that will benefit them throughout their lives. Remember that we are aiming for our children to have a growth mindset so that learning sticks.

Here are some simple strategies you can use in everyday life:

- **Break it into parts:** Don't overload them. Look at one skill at first, e.g. just focus on brainstorming. Next time, add pros and cons. Step by step, they'll learn the full cycle.
- **Modelling the process:** Show children how you approach and solve problems in your own life, whether it's a simple task or a more complex issue.
- **Talk through your thinking:** Talk through how you solve problems as they occur. This is easy to do as life unfolds.If you aren't sure what strategies you use, don't worry.We use different strategies for different problems in different areas.
- **Family problem-solving meetings:** Involve children in family problem-solving meetings, allowing them to see how you address challenges within the home.
- **Small challenges for them:** Encourage them to participate in solving smaller family problems, building their confidence and skills.
- **Ask guiding questions:** Use questions such as "What do you think will happen next?", "How could we solve this problem?", or "What other

options do we have?".

By incorporating these strategies when you are problem solving with your children, you can help them develop strong problem-solving skills that will benefit them throughout their lives. A growth mindset makes it stick.

Growth Mindset makes it stick.

A growth mindset helps children to see challenges as chances to learn. If you make problem solving fun, they'll want to do it more. That builds a **love of learning for life.** This priceless gift is one which you can give your children through simple activities in everyday life. They use problem solving skills all their lives and if you make it fun and challenging they will thrive when challenges are presented instead of feeling anxious and worried.

I love the song - "Try everything" *by Shakira* from the movie *Zootopia*. It has the best lyrics and is worth learning to sing with your children. Children love it from 4 - 12 year olds and beyond. The words give them an understanding that mistakes are part of the process of learning. Remember that we have said that our words become our children's inner voice. This is a great way of helping them to have a positive message about making mistakes in their heads on repeat. The song is very catchy and they all love singing it constantly which reinforces the message that everyone makes mistakes but you just get back up "to see what's next". I love the chorus words in the song which go "I won't give up, no I won't give in. Till I reach the end, and then I'll start again, No I won't leave, I wanna try everything, I wanna try even though I could fail"

Don't sweat the small stuff - problems to fix not disasters to fear.

Learning is all about making mistakes. It is important for children to know that they can try and fix any mistakes they make. It is also important that they understand what mistakes they can fix and what ones they can't. We once had a friend of my daughters over and he dropped a glass. It shattered all over the floor. He froze and didn't know what to do. He was looking around and obviously scared. My daughter was perplexed and asked him what he was doing. He whispered *"Won't your Mum get angry?"."As if"* she replied laughing " *wait here and I'll get the broom"*. That's when I walked into the room. Her laughter was contagious and they were both laughing with the glass all over the floor. My response was *"So, how do you want to clean this up?"* He was flabbergasted. I had just given them a problem to solve and there was no emotion. My daughter explained to him later that it was just an accident and you don't get into trouble for that. No need to be worried. He learned that accidents are just problems to fix, not disasters to fear. Taking the emotion out of the situation gave them the opportunity to solve the problem. The learning here is that you make your life easier when you **don't**

sweat the small stuff.

In addition it is also important to make sure you praise their effort and perseverance during the problem solving process, not just the outcome. That way, they learn that the process is also important.Finally, encourage them to keep going until they get the right solution. Sometimes it takes a few goes but that is completely normal and ok. They don't need to be perfect every time.

💡 Parenting Tip: Making learning sticky (and fun)

Here are some practical ways which you can help to make learning to problem solve fun.

Do it together. Choose problems where you genuinely don't mind the outcome—or can quietly fix later. If you routinely redo their efforts, they'll stop trying.

Use stories as rehearsals. Most children's shows/books follow a problem → attempts → solution. After an episode or chapter, ask:

"What was the problem?"

"What ideas did they try?"

"Why did the last one work?"

Kids then reenact the process in pretend play—perfect practice without real-world stakes.

Build with constraints. Once they've followed plenty of LEGO instructions, give them a picture of something and say, *"Make this with the bricks you have."* No purpose-made parts = rich problem solving.

Fun ways to support problem solving

🍰 **Snack-shop challenge:** Give a tiny budget and constraints (allergy-friendly, two colours on the plate). children plan, buy, and assemble.

- *Skills Built:* Budgeting, creative thinking, planning, and constraint-based design.
- *AI-World Connection:* Working within limits (budget, ingredients, colour rules) mirrors the way AI and engineers optimise for constraints in the real world. Children learn to balance resources, priorities, and creativity; skills essential for product design, sustainable business, UX design, and AI-driven logistics, where the challenge is to create smart, efficient, and ethical solutions under real-world limitations.

🎒 **Backpack pack-off:** List what must fit; children test layouts for speed/weight/comfort, then defend their design.

- *Skills Built:* Spatial awareness, efficiency, testing, and persuasive communication.
- *AI-World Connection:* Designing and justifying the best backpack layout introduces principles of optimisation, the same logic used by AI to solve problems like route planning or warehouse storage. Children practise defending their reasoning with data ("It's lighter this way"), a key skill for future roles in industrial design, AI modelling, robotics logistics, and engineering, where presenting data-backed solutions is critical.

Route race: Without the car, plan how to get to the footy using public transport or bikes. Time it!

- *Skills Built:* Planning, time management, navigation, and adaptability.
- *AI-World Connection:* Planning how to reach a destination without a car

teaches strategic problem-solving and systems thinking. It's the human version of how AI finds the fastest route using algorithms and data. Children learn to think ahead, adjust for variables, and test assumptions, preparing them for careers in AI transport systems, sustainability design, urban planning, and data analytics, where efficiency and adaptability shape smart cities.

Fix-it station: Keep a safe tub of broken-but-harmless items (old remotes/toys). Provide screwdrivers and a magnet tray; draw "how it works" sketches.

- *Skills Built:* Observation, engineering curiosity, hands-on exploration, and patience.
- *AI-World Connection:* Providing a safe space to tinker with broken objects helps children understand mechanical systems and cause–effect relationships. These cause and effect systems are the heart of both human invention and AI logic. This experimentation fosters resilience and curiosity, key traits for mechanical engineering, robotics, product testing, and AI hardware innovation, where learning through failure is the foundation of mastery.

LEGO constraints: Recreate a photo (bridge, animal, landmark) using only the bricks on the table; explain trade-offs.

- *Skills Built:* Creativity under pressure, visual reasoning, innovation, and communication.
- *AI-World Connection:* Recreating a design with limited pieces mirrors how AI systems work within data constraints. Children practise evaluating trade-offs, a vital part of real-world design and decision-making. These skills connect to architecture, design thinking, 3D modelling, and AI-generated creativity, where innovation often comes

from working within limitations.

What-if cards: Make a deck (e.g., "It's raining at recess," "Friend won't let you join," "Book Week costume ripped"). Draw, then run the loop.

- *Skills Built:* Emotional flexibility, empathy, decision-making, and social reasoning.
- *AI-World Connection:* Drawing unexpected scenarios ("It's raining," "Friend won't let you join") helps children practise adaptability and emotional problem-solving, the soft skills AI can't replicate. It trains them to pause, reflect, and respond with resilience. These human-centred traits are indispensable for leadership, counselling, negotiation, human–AI collaboration, and adaptive project management, where emotional intelligence drives creative outcomes.

AGE	ACTIVITY	EXAMPLE
1 – 3 years	Construction Toys	Wooden blocks, Plastic cups
	Obstacle courses	Make your own either in the house or in the back yard.
	Physical Challenges	Kindy gym, Jungle gym, Playground
	Stacking toys	Blocks, Jenga, Plastic cups
	Shape Puzzle ball	
	Push / Pull toys / Rocking toys	Cars, Rocking horse, Swing, Push bike, Ride on toys.
	Wooden Puzzles	Start simple
3 – 5 years	Board game	Storytime Chess game, Mancala
	Card games	Uno, Memory
	Construction toys	Duplo, Meccano, Magnetic squares, Mobilo, wooden blocks, k'nex
	Jigsaw puzzles	Start at 25 pieces and go up to 300
5 – 10 years	All of the above +	
	Board games	Chess, Ticket to ride, Catan, Rummy
	Jigsaw puzzles	300 + pieces
	Construction toys	Lego, Meccano, Mobilo, Magnetic squares
	Video Games	Limited Minecraft or Roblox **NOTE:** Please don't allow children to message or receive messages from others on this platform, even if they know them.
10 + years	All of the above +	
	Construction	Let's face it I still love Lego - you can start getting more complex and move into Robotic Lego
	Board Games	Monopoly, Cluedo, Pandemic, Risk, Crew
	Puzzles	Suduko, Logical Puzzles
	Video Games	Heavily supervised.

FINAL THOUGHTS ON INTERMEDIATE SKILLS

Now that the foundations are in place, this chapter turns curiosity into capability. We help children move from simply using information to testing, tracing, and improving it. The focus areas which we looked at were data protection & verification, general knowledge, research skills, understanding how things work, and practical problem solving. These skills use the earlier social emotional habits and give children the confidence to question AI, guide it, and spot when something doesn't make sense. The goal isn't memorising every fact; it's learning how to think like a checker, a maker, and a fixer.

Core Skills and Why They Matter

⊕ Data Protection & Verification

Data protection and verification cannot be understated. Children need to become trust detectors who are able to separate signal from noise in a fast AI-shaped internet. They learnt cheat codes for analysing all data in the form on SEFT-I (Source, Evidence, Freshness, Traceability, Intent). As well as practical tips on everyday safeguards such as privacy by default, strong logins, strong passwords, sensible backups, posting protocols, and the language to use in emails and posts.

In today's AI-driven world, children must learn to question what they see, read, and hear. Teaching data protection and verification skills helps them understand how to separate fact from fake, stay safe online, and value truth over popularity. These skills go beyond internet safety—they develop the mindset of a critical thinker who checks sources, looks for evidence, and considers intent before believing or sharing information. In the future, these same habits will serve them in careers such as data ethics, journalism, cybersecurity, and AI governance, where integrity and accuracy will define success. Future AI careers which Data ethics, Journalism, Cybersecurity analyst, AI governance, digital forensics, AI safety tester, misinformation researcher, trust & safety policy, compliance & audit.

🌏 General Knowledge (the "sense-making" layer)

General knowledge acts as a child's internal compass giving them context for what AI tells them. When children know how the world works, like how milk gets from the cow to the carton, how electricity powers their toys, or how history shapes today's news, they can instantly sense when something "doesn't sound right." This broad, interconnected understanding fuels curiosity and strengthens reasoning. In the AI age, strong general knowledge underpins innovation and critical thought. These skills are key to future industries such as education, creative industries, scientific research, and digital leadership. Careers which are at the heart of this are product manager, journalist/fact-checker, civic technologist, sustainability analyst, cultural liaison, education media creator.

🔍 Research Skills (from curiosity to credible answers)

Research is how curiosity matures into understanding. Teaching children how to ask good questions, find reliable sources, and check what they discover transforms them from passive consumers into active learners. They begin to see AI not as a shortcut, but as a tool to explore ideas more deeply and creatively. Children who master research will thrive in an AI world that values human originality and discernment—qualities at the heart of careers in innovation, design thinking, academia, policy, and creative problem-solving fields. Careers in which these skills will be needed are UX researcher, market/consumer insights, academic or R&D assistant, evidence-based policy, competitive intelligence, library & information science.

⚙ How Things Work (Systems Thinking)

Understanding how things work—mechanically, digitally, or socially—gives children a lifelong advantage in the future which AI is creating. It fuels curiosity, empowers independence, and builds confidence in tackling the unknown. When children can explain the process behind everyday systems, they become natural engineers, inventors, and thinkers. In a future where humans and machines collaborate, this skill will be vital for jobs in engineering, robotics, automation design, and sustainable technology development, where creativity meets technical insight. Careers which will benefit from these skills are process engineers, robotics/automation tech, operations & logistics, service design, IoT/smart-home technician, climate tech field roles.

Problem Solving

Problem solving is the bridge between knowing and doing. It teaches children how to stay calm when things go wrong, explore multiple solutions, and learn through trial and error. Every small challenge, from fixing a broken toy to navigating a friendship conflict, strengthens their ability to adapt, persevere, and think strategically. In the AI-powered workplace, these children will become the innovators who design solutions, lead teams, and improve systems. They will thrive in roles like project management, entrepreneurship, product design, and AI-human collaboration leadership. Future carriers for problem solvers are human-AI collaboration specialist, creative technologist, design thinker, startup founder, operations optimiser, product/solution architect.

Final Thoughts

Intermediate skills turn children into **thoughtful navigators** of an AI world, able to protect data, verify claims, ask sharper questions, and build workable solutions. These abilities map directly to fast-growing fields where human judgment, context, and creativity are the decisive edge.

INTERMEDIATE SKILLS AT A GLANCE

DATA PROTECTION & VERIFICATION

Children must become digital detectives - learning to separate truth from manipulation in an AI driven world.

AI World Advantage: AI Tools rely on data integrity. When children learn to assess credibility and dectect bias, They grow into responsible digital citizens who can question misimformation and strengthen the ethical use of technology

GENERAL KNOWLEDGE

General knowledge acts as a childs internal compass in a world overflowing with information.

AI World Advantage: AI tools are only as smart as their users. Children with strong general knowledge can challenge AI responces, decect flawed logic and connect ideas across disciplines, becoming the sense makers and innovators of the next generation.

RESEARCH SKILLS

Research turns curiosity into discovery. Teaching children how to ask thoughtful questions, seek credible sources and cross check what they find transforms them into lifelong learners.

AI World Advantage: AI can search but humans must make sense of what it finds. Research skills train children to guide AI, verify its outputs and generate integral insights an essential skill in innovation and information based industries.

HOW THINGS WORK

Understanding how things function weather mechanical, digital or natural - builds a mindset of curiosity and independence

AI World Advantage: Systems thinkers become the bridge between people and technology. They understand processes end-to-end, helping them spot inefficiencies and guide AI Applications in real world contexts such as robotics, automations and sustainability.

PROBLEM SOLVING

Analyzing issues and applying critical thinking skills

AI World Advantage: In an AI-augmented workplace, problem solvers are the innovators — they design solutions, refine systems, and collaborate with machines intelligently. They think beyond algorithms, applying creativity and empathy where AI cannot.

5

ADVANCED SKILLS IN DEPTH

With the foundations laid and intermediate skills in place, it's time to move into the advanced skills. These skills use the Foundation and Intermediate skills to learn new skills that will specifically set your child apart in the age of AI. These skills go beyond everyday learning and help children become not just capable users of technology, but confident creators, leaders, and innovators. The advanced skills include prompting, marketing and practical technical/technology skills. These skills are about integration: bringing together creativity, communication, and technology. They prepare children not only to thrive in school or their first job, but also to adapt, innovate, and lead in future careers we can't even imagine yet.

Just as in earlier chapters, you'll find tips, personal stories, and game tables to make learning fun and relatable. These are big topics, but remember that they don't need to be mastered all at once. Choose one or two to begin with, and layer on more as your child grows. Above all, keep in mind: advanced doesn't mean complicated. With practice and consistency, these skills become natural, giving your child confidence to use AI wisely, share their voice, and create their own opportunities in a fast-changing world. So let's begin this final stage where your child moves from learning skills, to using them to shape their own future.

✍ PROMPTING SKILLS AND CRITICAL THINKING

At its core, a prompt is simply a question or instruction you give to AI. But it's not just *any* question, it's one with enough detail and clarity that the AI knows exactly what you're asking for.

Think of a prompt as the recipe you hand over to a cook: if you just say, *"Make dinner,"* you might not like the result. But if you say, *"Make spaghetti bolognese with extra garlic and no cheese,"* you're far more likely to get what you want.

Lots of teachers have been using AI to create lesson plans and resources. When I first began experimenting, I asked AI to create a poem to create a song out of. I thought that I would be saving time, but after 30 mins of changing my prompts to try and get the right results I failed. Even then, I failed to prompt AI to formulate the poem I wanted with the detail I was asking. I reflected on my process. Was it my skills in AI prompting which was failing or was I just being picky about what I wanted? Truthfully it was probably both. I knew the type of result I was looking for, but AI just didn't come up with the goods. That moment taught me a powerful lesson:

the quality of the answer depends on the quality of the prompt.

Why detail matters

Prompting is all about precision and detail. The more specific and descriptive you are, the better the outcome. Children who practise giving clear, detailed instructions are unknowingly training themselves to become great prompters in the AI world. One of the best ways to do this is through storytelling and procedure writing. These activities help children practise breaking down their thoughts step by step, while also teaching them to use adjectives and detail in everyday communication.

🔅**Parenting Tip:** Learning to prompt AI can be as easy as making your children be specific when asking for anything. "Can you please pass the white salt container?" instead of "Pass that please". Actually it has been my children who have taught me how to specifically ask for what I want. They love playing tricks on me if I am not specific enough. It's been a game changer for me and has inadvertently helped me work with AI.

Fun ways to teach prompting skills

1. Procedure writing (from age 5+)

Ask children to explain a simple task step by step (e.g. making a sprinkle sandwich) and then you do exactly what they say.

At first, they might say, *"Put the sprinkles on the bread."* You then place the entire jar on the bread (cue giggles)! They quickly realise they need to give clearer instructions: *"Open the jar, pour the sprinkles onto the bread, then spread them out with a knife."*

This playful misunderstanding helps children refine their instructions until they're accurate and detailed. It's also lots of fun for us adults when the kids roll their eyes and say *"Not again Mum, you know what I mean"* and I reply *"Well be more specific."*

2. Practise adjectives and description

Encourage kids to use describing words whenever they make a request. For example: Instead of *"I want strawberries,"* guide them to say: *"I'd like ripe strawberries cut in half on my dessert, please."*or have them recall and explain a recipe they've helped with, step by step, for a friend.

🔅**Parenting Tip:** Create a craft procedure card collection: let children describe the steps for a craft project, then you write it down for them. Later, they can follow the card to recreate the project. This reinforces the importance of clear, descriptive language.

3. Be specific in requests

AI struggles with vague instructions, and frankly so do parents! If your child asks for a *"glass,"* hand them an empty glass. When they look puzzled, explain: *"You didn't say a glass of water, you just said a glass."*

At first it might frustrate them, but soon they'll learn to say exactly what they mean. For example: *"Can I please have a glass of cold water?"*

This skill of asking specifically mirrors the precision they'll need when prompting AI to produce reliable results.

4. Use puns and humour

Kids love jokes and wordplay. If they ask for something vaguely, playfully give them the wrong thing on purpose. Their laughter makes the lesson stick: being clear and detailed avoids confusion.

Why this skill matters for the AI era

Prompting isn't just about AI, it's about communication, clarity, and problem-solving. By teaching children how to be specific in their requests, use adjectives, and break tasks into steps, you're equipping them with one of the most important life skills for the AI-driven future. If you aren't specific enough it could use false data or give you something completely different to what you want. Teaching children how to ask for things with detail is as simple as incorporating detail into everyday life.

The better they are at asking the right questions, the better their results will be, whether from AI, from school projects, or in real-life teamwork.

Fun ways to support Prompting skills

⬡**Sprinkle Sandwich Challenge:** Write out each child's steps and then literally follow them, even if it means chaos and a big mess.

- *Skills Built:* Clarity in communication, sequencing, logic, and attention to detail.
- *AI-World Connection:* Writing and following imperfect instructions shows children how vague prompts create confusing results — exactly what happens when humans give unclear directions to AI. By refining their wording, they practise the essential AI-era skill of prompt engineering. This simple, messy game prepares them for future roles in AI design, robotics control, digital instruction, and workflow optimisation, where precision in language drives successful outcomes.

⬢**Recipe Recall:** Have them explain how to make pancakes, spaghetti, or their favourite snack without looking at a recipe.

- *Skills Built:* Memory, structured thinking, and verbal reasoning.
- *AI-World Connection:* Explaining a recipe from memory encourages children to think step-by-step, teaching them how to structure processes which is the same way AI systems break problems into smaller tasks. This builds cognitive skills valuable in programming, instructional design, systems analysis, and AI-assisted education, where clarity of process and logical sequencing are key to getting the right results.

🎨 **Craft Cards:** Create step-by-step guides for art projects, then swap them between siblings or friends to see if the instructions make sense.

- *Skills Built:* Procedural writing, teamwork, feedback, and iteration.

- *AI-World Connection:* Swapping written instructions for an art project mimics the feedback loop between humans and AI. When one person's "prompt" produces unexpected results, both must refine the wording and re-test. This teaches iterative problem-solving, an essential skill for UX design, human–AI communication, technical documentation, and collaborative innovation, where clarity and revision improve outcomes.

♣ **Detail Detective:** One child gives a vague request; the parent plays "AI" and deliberately misinterprets. The child must refine their request until it's crystal clear.

- *Skills Built:* Observation, precision, problem-solving, and metacognition.
- *AI-World Connection:* When a child refines a vague request after the parent (acting as "AI") misinterprets it, they're learning exactly how humans must think when prompting digital tools. It builds awareness of how language shapes output and trains analytical patience which is ideal for careers in AI prompt engineering, quality assurance, computational linguistics, and customer experience design, where refining instructions is both an art and a science.

Adjective Shopping List: Write a list for groceries together, but insist on including adjectives ("crispy apples," "ripe bananas," "soft bread").

- *Skills Built:* Descriptive language, categorisation, and sensory awareness.
- *AI-World Connection:* Adding adjectives ("crispy apples," "soft bread") helps children learn that words add context, just as detailed prompts help AI produce relevant answers. They see that specificity saves time and reduces confusion which is a key communication principle in marketing, content creation, AI writing, and data labelling, where rich, accurate

descriptions produce better results and stronger human–machine collaboration.

AGE	ACTIVITY	EXAMPLE
1 – 5 years	Word knowledge and play	Vocabulary – From the beginning please use the correct words for items. No baby talk. Describe items while walking around
5 – 10 years	Puzzles	Crossword puzzles, Wordsearch
	Board Games	Scrabble, Code names, The Chameleon, Herd Mentality
10 + years	AI apps for finding information with parents.	Working with AI apps to find information out with parental supervision

✎ MARKETING SKILLS AND STRATEGIES

Marketing skills is the ability to use creativity, communication, strategic thinking and data analysis to promote something. Traditionally, this meant selling products or ideas. But in an AI-driven world, marketing is also about **how we present ourselves**. Marketing skills require good strategies that provide plans on how to convince people that you have the right solutions. AI can perform many technical marketing tasks like writing ads, analysing customer behaviour, or even designing graphics. But it cannot replace the human ability to connect, build trust, and tell a story.

Marketing skills are useful to our children for their future as having great job skills isn't always enough. It's about helping them to understand their strengths, present their ideas confidently, and share their talents with the world so they don't get lost in the noise. When they enter any workforce they need confidence, social skills and knowledge but they will ultimately need to market their ability to work and communicate their value. How do you describe what you have to offer companies or others? Why should they employ you when AI can do most things for them? It's about knowing what to market and looking at trends so that you can arrive at the right solution. AI is not flexible enough yet to see outside the box and innovate. AI can teach you but you need to know that marketing exists. You need to know the general aspects of marketing and how it applies to you and what you are doing. This is the era of the generalist. A jack of all trades. Understanding generally what is needed, but using AI to get the details.

Why Marketing Matters More Than Ever

In the past, people could quietly work hard and expect to be noticed. Today, that's not quite enough. Algorithms, AI-driven hiring systems, and global competition mean that opportunities often go to those who stand out and communicate well. Marketing skills empower your child to:

- Get noticed in school, sports, and creative pursuits.
- Build a personal brand that reflects who they are and what they value.
- Adapt to future job markets, where AI may handle technical tasks, but humans will still need to connect emotionally and creatively.
- Take ownership of their story, instead of letting others define them.

Think of these skills as a *superpower* which gives them the ability to make their ideas shine and open doors to new opportunities.

The 6 Steps to Marketing Yourself

Our children need a guide to presenting themselves in the AI era. As they saying goes;

"It's about who you know and not what you know"

This saying is even more relevant in the future AI world than it has been in the past or even today. It's about making those human connections and finding your tribe. There are 6 basic steps we can help our children go through to learn how to market themselves. You might like to use these as well to improve your job advancement chances. They work for everyone.

Step 1: Start with Self-Awareness

Before your child can market themselves, they need to understand who they are and what they bring to the table. This begins with reflection:

Strength spotting: Ask them, *"What are three things you're really good at?"* or *"What do your friends or teachers say you do well?"*

Values check: Discuss what's most important to them, kindness, creativity, fairness, problem-solving and how those values shape their choices.

Passion projects: Encourage them to explore hobbies and interests. The things they're naturally drawn to will often point toward their unique "brand."

> 💡 **Parenting Tip:** Create a "strength map" together, drawing out words and images that represent their skills and passions. This visual can grow as they do.

Step 2: Teach the Basics of Storytelling

Marketing is really about telling a compelling story. Whether your child is introducing themselves to a new teacher or pitching a project idea, a clear story will help them connect with others. They can ask the following questions:

- Who I am: *"I'm Sarah, and I love solving puzzles and building things.",*
- What I do well: *"I recently performed my violin by myself in a concert.",*
- Why it matters: *"I love to perform for people and make them happy"*

> 💡 **Parenting Tip:** Practice this in fun, low-pressure settings. For example, have your child "pitch" themselves to grandparents at dinner, or introduce themselves at family gatherings using this framework.

Step 3: Digital Presence in the AI Age

AI is shaping the online world, from how recruiters find candidates to how search engines display information. Even children need to learn digital self-marketing basics, in safe and age-appropriate ways. Here are some ways we can guide our children as they go from 6 years through the teens.

Privacy first: Always discuss boundaries relating to what to share, what to keep private, and how to protect personal information. For younger children, focus more on understanding safe digital behaviour than building an online profile.

Social media literacy: They should not have social media until they are 16 years old. I can remember when one of my daughters got social media.She posted a picture which was not the best and her older sister ran down the stairs yelling at her to get that picture down. She had her back and then we were able to have the conversation about their digital footprint. Teach them to treat every picture and post as part of their "digital footprint" which will be looked at years into the future when they go for jobs.

Portfolio creation: Encourage older children or teens to build a simple online portfolio showcasing projects, art, writing, or achievements. Websites like Canva and Wix make this easy.

LinkedIn for teens: Some teens create profiles to start networking early. Guide them to focus on skills and values, not just achievements.

AI tools for visibility: Introduce them to beginner-friendly AI tools that help with creating resumes, designing graphics, or even practicing interview questions.

Step 4: Practice Elevator Pitches

An elevator pitch is a 30-second introduction that quickly communicates who you are and what you do, taking the information we did in step 1 and presenting it to different people. This is a fun skill to practice and can grow with your child over time.

💡 **Parenting Tip:** Turn this into a family game: Everyone writes their pitch on a card and shares it. Vote on the most creative, clear, or inspiring pitch. Offer positive feedback and ways to improve.As an added bonus, this helps them to learn how to receive feedback in a healthy positive way.

Step 5: Encourage Collaboration and Networking

Marketing isn't just about promoting yourself, it's also about building relationships and collaborating with others. We have discussed collaboration in foundation skills. Use these techniques to teach your child to be a great collaborator and networker by;

- **Show Gratitude:** Saying thank you with a handwritten note or thoughtful personal message;
- **Helping:** Offering help and encouraging children to support their friends and classmates, not just seeking help from others;
- **Connecting ideas in conversation**: When they see someone with similar interests, prompt them to start a conversation and collaboration. For example *"I saw you're into robotics too, want to work on a project together?"*

These skills will prepare them to thrive in both school group projects and future professional networks.

Step 6: Teach Ethical Marketing

The AI world is filled with ads, influencers, and clickbait. It's critical to teach children that authenticity and ethics matter. This helps them grow into leaders who are both successful and trusted. Here are some important ethical issues to consider;

- Marketing should tell the truth, not manipulate.
- They should own their mistakes and take responsibility for their message.
- Encourage them to think about how their words and actions affect others.
- Discuss real-world examples of misleading advertising or deepfakes to help them spot unethical behaviour.

💡 **Parenting Tip:** Show your children an age appropriate deepfake video. There are plenty around. Start a conversation about *"How do we know what's real? What responsibility should the creator have?*

Are there any laws which prevent this? Who should we report it to?"

The Goal: Confidence, Not a brand

The aim of teaching marketing skills isn't to turn your child into a brand at a young age, it's to give them confidence and clarity. When they can talk about their ideas, share their talents, and connect with others, they'll feel empowered to create their own path. Though AI may be reshaping the world, human stories, creativity, and relationships will always matter. By giving your child these tools now, you're preparing them not just to *fit into* the future, but to stand out and shape it. Marketing yourself is important however in the real world it is important to be able to market in the AI-powered marketplace.

Future AI-powered Marketing Skills

The rise of artificial intelligence is reshaping how products, ideas, and even people are marketed. In the future, many traditional marketing tasks will be handled by AI, from writing ads to analysing customer data. This doesn't mean humans won't play a role. In fact, human creativity, empathy, and strategic thinking will be more valuable than ever. For your child to succeed in this new world, they'll need to work alongside AI, using it as a tool rather than being replaced by it. This chapter has been working through helping your child develop future-ready marketing skills so they can lead, not just follow, in a marketplace transformed by technology.

AI has already made significant changing to marketing throughout the world. AI is already behind many things we see every day; personalised ads on social media, chatbots answering questions on websites, algorithms recommending videos, songs, or products, Tools that design logos or write marketing copy. In the future, AI will become even more advanced. Instead of just assisting marketers, it will drive marketing campaigns from start to finish. This means human marketers won't need to do repetitive tasks, like sorting data or scheduling ads. Instead, their role will be to guide the AI, tell compelling stories, and make ethical decisions.

"AI can process information. Humans must provide imagination and integrity."

In the future humans will be providing the creativity, making ethical decision, interpreting data with empathy and context and collaborating with AI.

The Four Core Marketing Skills

These Four Core Marketing Skills are the over riding skills which children need to market any product. They are slightly different to the 6 Steps of Marketing Yourself as the 6 steps focuses on yourself and the 4 Core Marketing Skills focuses on marketing in general.

The 4 Core Marketing Skills are; Data Literacy, Human Creativity, Ethical Decision Making and Collaboration with AI. To thrive in an AI-driven marketplace, children will need to master all four key skill areas. These can be introduced gradually through fun, and age-appropriate activities.

Data Literacy – Understanding the Numbers:

Your children will have learned all about data integrity and verification in the intermediate skills. We all know that AI runs on data.It collects and analyses information we ask it to, from information about customers habits, to products sold, and trends in sales. Future marketers will need to understand how to read, interpret, and question data rather than just collect it. Key concepts to introduce to our children are:

- **Patterns:** Teach your child to notice patterns in everyday life. "Which day of the week do most people visit the playground?"
- **Surveys and feedback:** Have them design a simple family or friend survey, then chart the results.
- **AI and bias:** Discuss how data can be *incomplete or unfair.* If a survey only asks one group of people, the results don't show the full picture.

🔅 **Parenting Tip:** Create a "data detective game." Collect basic data (like family snack preferences) and have your child analyse it to decide what to "market" at a pretend snack shop.

Human Creativity – The Un-replaceable Skill:

AI can write a catchy slogan, tell a story or design a logo in seconds, but true creativity still comes from humans. AI can generate ideas, but humans provide the imagination to make those ideas meaningful.

🔅 **Parenting Tip:** Here are some great ideas to do at home; **Storytelling Challenges:** Give them a random object and have them create a 30-second ad for it. **Idea Mash-ups:** Combine two unrelated things (like a skateboard and a backpack) and imagine a product that uses both. **Content Creation:** Encourage them to make simple videos, podcasts, or posters around causes they care about.

Ethical Decision-Making – Doing the Right Thing:

The future of marketing will involve tough ethical questions. AI can create deepfakes, manipulate emotions, and spread misinformation at lightning speed. Children must learn how to use these tools responsibly. By teaching ethics early, you'll raise a child who values integrity over shortcuts, even in a competitive AI-driven world.

> 💡 **Parenting Tip:** Discuss these real-world issues over the dinner table like: Why honesty matters in advertising? How targeting the wrong audience can cause harm? The importance of respecting privacy and consent? The difference between persuasion and manipulation?

Collaboration with AI – Becoming the "Human in the Loop"

In the future, AI will be like a powerful assistant. As with any assistants today, we will have to guide it, check its work for mistakes, bias or missing information, and add our own touch. Children must learn to collaborate not only with other people, but also with AI tools, treating them as partners rather than competitors. AI can process, but humans must interpret.

> 💡 **Parenting Tip:** Here are 3 practical ways you can help your child to become the best collaborator with AI. **Prompt writing:** Teach your child to see the detail as we discussed in an earlier chapter. As they get older, show your child how a well-written prompt leads to better AI-generated content. **AI as a brainstorming buddy:** Use an AI tool to generate ideas, then have your child pick the best and improve them. **Human oversight:** Practice checking AI outputs for mistakes, bias, or missing information.

The Human Advantage

As AI continues to evolve, it will take over many technical marketing tasks. But there's one thing it can never replace: the human heart and imagination.The marketers of the future won't just push products — they'll build trust, shape culture, and solve problems in ethical, creative ways. By teaching your child these skills now, you're preparing them to lead the conversation, not just follow the algorithm.Encourage them to see AI not as a threat, but as a tool that amplifies their unique human gifts. When children grow up understanding how to use AI thoughtfully, they'll be ready to thrive in a world full of both challenges and possibilities.

Fun ways to support Marketing

📺 TV Shows (The Gruen Transfer) - As my children became teenagers we loved watching a show called 'The Gruen Transfer' on the ABC, which gave insights to marketing and how it works. They really learnt so much from this show and got a good laugh as well. We had interesting and meaningful discussions during and after each episode as a family. The following activities are designed to build AI-era marketing skills through fun, hands-on experiences.

- *Skills Built:* Creativity, marketing, general knowledge.
- *AI-World Connection:* Shows like The Gruen Transfer help children see how influence, storytelling, and psychology shape consumer decisions all critical in a world where AI analyses human behaviour to sell products. By discussing what makes ads persuasive or manipulative, children develop media literacy and ethical awareness which is essential for future careers in digital marketing, behavioural science, AI content strategy, and ethics in advertising.

 Family "Product Launch" Game- Have your child invent a product (real or imaginary). Use an AI tool to design a logo or slogan. Your child creates the pitch and story behind the product. Family members vote on the most convincing campaign.

- **Skills built:** Creativity, storytelling, collaboration with AI.
- **AI-World Connection:** This activity introduces children to human–AI collaboration in creativity. When they use an AI tool to generate slogans or design logos, they're learning how to guide technology with clear prompts which are the foundation of tomorrow's creative AI work. These are the same skills used by brand strategists, product designers, entrepreneurs, and AI marketing specialists. The fun of pitching ideas also strengthens communication, innovation, and adaptability which are timeless success traits.

⚖ "Good vs. Misleading" Ad Challenge - Look at real ads together (from TV or online). Identify which are honest and which are manipulative. Create an "ethical ad" for a similar product.

- **Skills built:** Critical thinking, ethics, empathy.
- **AI-World Connection:** In an age where algorithms and AI-driven campaigns can distort truth, being able to spot manipulation is a superpower. This game helps children think critically about ethics in data-driven advertising. They learn that technology can persuade, but humans must choose integrity. These insights prepare them for roles in ethical AI policy, consumer protection, and responsible marketing, where empathy and transparency are the new currency.

🤖 AI-Powered Brainstorm Session - Choose a topic, like saving the environment. Ask an AI tool for five creative campaign ideas. Your child picks their favourite and develops it further. Discuss how AI's ideas compared to their own.

- ***Skills built:*** Prompting, human-AI collaboration, problem solving.
- ***AI-World Connection:*** This challenge transforms children into AI co-creators. By prompting an AI to brainstorm campaign ideas and comparing the output to their own, they learn how to evaluate, refine, and direct machine-generated creativity. This builds prompt engineering, analytical thinking, and creative leadership — skills used by future AI campaign managers, innovation consultants, and sustainable brand developers who know how to harness AI rather than be replaced by it.

🌐 Marketing "Mission Statement": Help your child write a one-sentence mission that represents their values as a future marketer or creator. Examples are "I use technology to tell stories that inspire kindness."; "I want to create honest ads that help people make good choices."

- ***Skills built:*** Self-awareness, clarity, goal setting.
- ***AI-World Connection:*** In the AI era, where technology can amplify messages faster than ever, purpose and ethics matter more than profit. Helping a child define their marketing mission cultivates values-driven leadership, an anchor in a rapidly changing world. These reflective skills lead to future opportunities in social entrepreneurship, digital storytelling, brand ethics, and corporate social responsibility, where integrity and authenticity set human creators apart from AI-generated noise.

Create a Mini Business – Help your child design and "sell" a simple product or service (like handmade bookmarks).

- ***Skills Built:*** Entrepreneurial thinking, creativity, communication, financial literacy, problem-solving, and resilience.

- *AI-world connection:* Children learn how to identify needs, create value, and communicate that value — the same skills used by innovators, social entrepreneurs, and creators in the digital economy. Future roles include product designers, digital entrepreneurs, ethical marketers, and AI startup founders.

🎨 Design a poster or ad – Use Canva to make a digital flyer for their favorite hobby or club.

- *Skills Built:* Digital design literacy, visual communication, creative marketing, and audience awareness.
- *AI-world connection:* This activity teaches children to blend technology and storytelling — crucial in careers like digital media, user experience (UX) design, creative direction, and AI-assisted content creation. They learn how design influences emotion, decision-making, and trust — skills AI cannot replicate authentically.

💬 Host a mock interview – Pretend to be a hiring manager and ask questions about their skills and goals.

- *Skills Built:* Self-awareness, confidence, articulation, emotional intelligence, and adaptability.
- *AI-world connection:* Interview practice builds human-centered communication, a key differentiator in careers that rely on empathy and clarity, such as leadership, education, consulting, and AI ethics liaison roles. It helps children learn how to express their value, something machines cannot do.

Run a "Shark Tank" night – Have family members pitch creative project ideas, and everyone "invests" using fake money.

- ***Skills Built:*** Critical thinking, persuasion, teamwork, and creative problem-solving under pressure.
- ***AI-world connection:*** Children learn to pitch ideas, receive feedback, and refine, mirroring innovation cycles in tech startups, AI product design, and sustainable entrepreneurship. It's about learning to fail forward and think like an innovator, adaptable, articulate, and resilient.

🌍 Make a values-based mission statement – Guide them to write one sentence that describes what they stand for and how they want to make a difference.

- ***Skills Built:*** Ethical reasoning, goal setting, self-reflection, and clarity of purpose.
- ***AI-world connection:*** Defining values is essential in a world where AI lacks moral compass. This activity nurtures ethical awareness — a foundation for careers in AI policy, digital ethics, psychology, and community leadership. It helps children align their choices with who they want to become.

AGE	ACTIVITY	EXAMPLE
1 – 5 years	From 3 look for logos	Find a MacDonalds sign, Nike,
	Positive re-enforcement of good qualities they possess	Your voice becomes their inner voice and their reality. Say "I love how you solved that problem, you are the best at problem solving" "I love how quickly you got ready, you are becoming so independent" "That was really kind when you helped your friend. You are so kind"
5 – 10 years	Board Games	Ticket to ride, Catan
	Talk about tag lines and logos	Hungry Jacks "The burgers are better at Hungry Jacks"
	Design their own logo for their name	
	Make a restaurant at home	Children choose the menu, logo and marketing the restaurant name etc and have grandparents or friends over. They can make up menu's and have a drinks menu etc.
	Build vocabulary	Increase their knowledge of adjectives (describing nouns / things) and adverbs (describing verbs / actions)
10 + years	TV Shows	Gruen Transfer show on ABC (make sure you watch the episode first to make sure it is appropriate for the age group)
	Own business with supervision	Create a gardening or cleaning company which the children run with your supervision of course.

🖥 PRACTICAL TECHNICAL SKILLS

We live in a world where technical skills are no longer optional, they're essential. But before diving into what children should learn, we must balance skill-building with healthy screen use. Remember we spoke about children under 5 not using screens for more than 30 mins per day. That goes up to 1 hour for primary school kids. The research by Muppala, Vuppalapati, and Pulliahgaru (2023) found that while some believe screen media helps with learning, in reality it often has **negative effects** on children's development. High screen time can impair executive functioning, sensorimotor development, and academic outcomes later in life. It is also linked to social-emotional difficulties, sleep problems, and anxiety.

So how do we find that balance during school age and still learn all the skills we need? It is hard, but it is about prioritising what they are allowed to use their screens for. Finding opportunities to practise useful digital skills, while still protecting their brains, emotions, and resilience.

Essential Tech Skills for Children

These are a few of the technology skills all children need to be able to do. Schools are increasingly asking parent to Bring Your Own Device to school. It is essential that your child understands how to safely use the device and that is up to you as a parent to oversee.

Touch Typing helps students speed up their communication and reduces mistakes when interacting with AI.

Social Media Awareness is paramount in this day and age, with new laws prohibiting social media giants from students under 14. What ever they put up online, stays online to be accessed well down the track.

File Organisation and Saving Document protocol is about where we put our files and what we call them.

Digital communication etiquette is all about how we write emails and have the right tone in our online writing

Basic cybersecurity is essential as children need to know how to create strong passwords, know how to keep safe online by never sharing personal details or click unknown links

Search and verification skills like using multiple sources and understanding bias in sources

Safe use of AI Tools is looking at verifying answers, writing clear prompts and teaching students how to work with AI.

Basic troubleshooting when tech doesn't work. Have you every heard, just turn it off and on again? You try it and it works. Help them to explore the setting menus and the basic fixes for all technology.

Touch Typing

One of the most valuable skills for the AI era is **touch typing**. It speeds up communication, reduces mistakes, and makes working with AI tools smoother.

My dad insisted I learn touch typing in school, and it has been a lifelong advantage. Sadly, many schools today no longer teach it. Parents can give their children an edge by encouraging them to practise with typing games on a laptop or a proper keyboard attachment for tablets.

I hear you say that your children use voice-to-text all the time. I'm not denying that it is handy, but it still requires editing and keyboard fluency. Knowing the keyboard makes corrections fast and efficient.

> 💡 Parenting Tip: Use a **real keyboard**, not just a touchscreen. Encourage **no peeking** by covering hands with a cloth while they practise. Make it fun with **typing games** or races. **Repetition is the key:** Encourage short, daily practice. This daily practice develops the muscle memory and makes it automatic.

🔆 Parenting Tip: If you don't touch type, then maybe you should join them and learn the benefits first hand.

Social Media

The safest rule for teens: **delay social media as long as possible** (ideally until 16+). Why? Because every photo, post, and comment creates a digital footprint that can follow them for life. Employers, universities, and organisations check these footprints when making decisions.

A friend of mine who hires staff always reviews candidates' online presence. If two applicants look equal on paper, he chooses the one with the stronger digital footprint.

Teen's brains are not yet wired for the pressure of high-stakes digital spaces. They are not equipped to make good decisions about what they use social media for. Even the best teen can make one poor decision which can cause long-term consequences. One of my daughters once posted a rash photo on Snapchat. Her sister saw it within seconds and I heard her running down the stairs yelling, *"delete that photo now."* Thankfully her sister caught it and they deleted it before it left a mark. That quick intervention prevented a lasting mistake, but not every teen has that safety net or is as lucky. As the saying goes - It's better to be safe than sorry.

🔆 **Parenting Tip:** Teach them to think, if in doubt, don't post.

File Organisation and Saving document

How many times have you lost your files and can't find them because you have either not named them properly or put them in a folder somewhere. Losing a file is frustrating and can lead towards many hours of hard work gone. Teach your children to be organised with their files on their devices. Saving file protocol and file organisation can be the difference between losing the work you have done over the last 8 weeks or just losing a few hours work. Making sure you name your files in ways which are easy to locate later is a skill in itself. Help them to organise their files into folders which make sense. These are skills which you can help your children to master but are not necessarily taught elsewhere, makes life so much easier for them at school and beyond.

Good File Organisation and Saving habits to teach children:

- **Naming files clearly:** *Science_Project_Term2.docx* is better than *FinalFINALv2.docx, or presentation 8.pptx*
- **Folder systems:** Use folders to group files by subject or project. Put all the files together for one project whether it be photos or files.
- **Backups:** Save to the cloud or an external drive regularly. Also make new versions so that you have a few versions to go back to if your work gets lost or corrupt. This has saved me many times e.g. TechPresentatio nAugv1 TechPresentationAugv2 etc.

This skill saves hours of stress in school and beyond.

4. Digital Communication Etiquette

AI can generate text, but **how you communicate still matters.** There are three things which go into a good communication;

- How to write clear and polite. Emails and messages use different etiquette. What is good for one is not necessarily polite in another form.
- To think about **tone.** This is difficult to explain to your child, so it's about understanding that words can be misread online. Make sure your communications cannot be read a different way and people could take offence.
- To avoid oversharing personal details. Make sure they understand the difference between chat groups and real life. You don't know who is on the other end of the chat.

💡 **Parenting Tip:** Have them draft an email to a teacher or coach. Review it together and talk about tone, clarity, and respect.

5. Basic Cybersecurity

This is a big one too. There are many predators online. We cannot get away from this fact and we need to build in safety procedures in our daily lives online to help prevent them accessing our children. Although I don't advocate for you to talk about predators with the children, they need to understand that not everything online is safe. Teach them how to keep themselves safe through these key lessons:

- Create **strong passwords** and never reuse them.
- Don't click on unknown links or attachments.

- Never share personal details (address, phone, passwords) online.

💡 **Parenting Tip:** Play "phishing detective"—show them mock messages and let them spot which ones are scams.

6. Search and Verification Skills

Having the internet has transformed our access to information over the last few decades. AI tools are now providing the next level of information dissemination. AI can provide answers, but children must learn how to **check if those answers make sense.** Here are some quick ways to help children to analyse the answers which are given when seeking information on the internet or with AI tools.

- Use multiple sources, not just one.
- Ask: *"Does this information align with what I already know?"*
- Understand bias in sources (ads, opinion sites vs. factual reporting).

💡 Parenting Tip: Let them research a topic of interest (e.g., "Who invented soccer?"). Then compare answers from Google, a book, and AI. Discuss differences.

7. Safe Use of AI Tools

As children begin experimenting with AI (chatbots, art generators, homework helpers), teach them:

- How to write clear prompts, see earlier in this chapter for ideas on how to do this.

- To double-check AI answers against trusted sources.
- Why it's wrong to pass off AI's work as their own.

This builds responsibility while making AI a tool, not a crutch.

8. Basic Troubleshooting

Children often panic when tech doesn't work. Some children are better than us adults in determining what is wrong with their device. You can build your childrens' confidence in troubleshooting with small steps which are the underlining basis of troubleshooting:

- Restart the device.
- Check Wi-Fi or battery.
- Explore settings menus for simple fixes.

💡 **Parenting Tip:** Next time something goes wrong with the TV or printer, walk your child through the fix instead of doing it for them.

Fun ways to support Basic Tech skills

⌨ **Typing races:** Compete to finish short texts quickly and accurately.

- *Skills Built:* Speed, precision, focus, and hand–eye coordination in digital environments.
- *AI-World Connection:* Typing fluency helps children communicate seamlessly with technology, an essential gateway skill for prompt writing, coding, and creating digital content. These micro-skills

underpin roles in AI operations, digital communication, and data input automation, where speed and clarity of thought matter.

File hunt challenge: Hide a document and challenge your child to find it. Make it more of a challenge by timing them.

- **Skills Built:** Digital navigation, logical reasoning, organisation, and persistence.
- **AI-World Connection:** Understanding how to locate and manage files mimics *data retrieval* — a foundational concept behind AI systems. This skill leads toward data management, cybersecurity, and information architecture, where knowing where things "live" digitally is half the job.

Digital spring-clean: Sit together and "spring clean" your devices. Organising files together regularly.

- **Skills Built:** File organisation, data hygiene, and responsibility for digital clutter.
- **AI-World Connection:** When children learn to maintain "clean" data, they understand one of AI's most important rules: garbage in, garbage out. This builds habits valued in data science, cloud management, and digital archiving, where accuracy and structure are key.

Phishing quiz: Print fake vs real emails for them to spot the difference.

- **Skills Built:** Cyber awareness, pattern recognition, and critical judgment.
- **AI-World Connection:** Spotting fake links or messages teaches *threat detection* and *data protection* — vital in a world where AI systems are

trained to flag fraud and misinformation. These habits connect to future paths in cybersecurity, AI trust and safety, and digital forensics.

🤖 **AI sandbox:** Safely experiment with AI tools, but always reflect afterwards: *"Did this answer make sense?"*

- *Skills Built:* Safe experimentation, prompting, and critical reflection.
- *AI-World Connection:* By testing AI tools and analysing their answers, children learn *human–AI collaboration.* They discover that machines can help — but humans must verify. These thinking skills map directly to roles in **AI design, machine learning ethics, innovation consulting, and education technology.**

🧠 **Fix-it together:** Let them be your "assistant" when solving a tech issue at home.

- *Skills Built:* Troubleshooting, resilience, curiosity, and systems thinking.
- *AI-World Connection:* When children help fix a Wi-Fi issue or a printer problem, they're learning diagnostic reasoning. This is the same mindset used by engineers and AI analysts to debug complex systems. This nurtures future potential in IT support, robotics maintenance, and AI engineering.

📄 **Mock interviews:** Role-play as an employer checking someone's "digital footprint." Let your child think through what they'd want found online.

- *Skills Built:* Digital citizenship, reflection, and reputation management.
- *AI-World Connection:* Understanding one's online presence builds awareness of digital identity which is a vital soft skill for a career world

shaped by AI hiring systems and online profiling. This awareness links to careers in personal branding, HR technology, and digital leadership, where integrity meets visibility.

AGE	ACTIVITY	EXAMPLE
1 – 5 years	No tech	
	Play outside	Playground, park, beach,
	Pretend Play	Build a Fort, Play with a dolls house, Make a cubby, Pretend tool kits, Toy kitchens, Toy cleaning equipment
	Fine motor games with fingers	
	Action songs	eg. Incy wincy song with finger actions
5 – 10 years	iPad / tablet introduced	No Phone or messaging unless heavily supervised by parents and parental controls.
	Touch typing	Online games to learn how to touch type
	Educational games	Scratch, Coding Safari, Reading Eggs, Mathletics, National Geographic children.
	Book creator	An app to make your own books with photos and words
	Powerpoint / Excel / Word apps	Word processing, presenting and number apps which are used throughout their lives. Create own docs to share within the family.
	File naming conventions and saving techniques	How to create folders and sort your documents. Name your document with the year, month and what it is at least.
10 + years	As above +	
	No Social Media until 16 years.	This may seem harsh but they are creating a digital footprint from day 1.
	Phone etiquette and rules	What to share and what not to.
	Video games	

FINAL THOUGHTS ON ADVANCED SKILLS

By the time your child reaches the advanced stage of *The Advantage Code*, they're no longer just learning how to use technology, they're learning how to lead it. These skills prepare teenagers and young adults to become thoughtful creators, innovators, and ethical decision-makers in an AI-driven world. The goal now is to help them direct technology, question outcomes, and build purpose-driven influence.

This chapter focuses on three core areas: Prompting and Critical Thinking, Marketing Skills and Strategies, and Practical Technology Skills. Together, these form the bridge between human intuition and digital intelligence, empowering young people to shape the future rather than be shaped by it.

Core Skills and Why they matter

🔍 Prompting and Critical Thinking

AI may hold endless data, but it still depends on the clarity and curiosity of the human asking the question. Prompting and critical thinking teach children to communicate precisely, evaluate results, and separate useful information from noise. They learn how to *guide* AI systems, not simply accept their answers. In the future, success will belong to those who can think creatively, challenge assumptions, and design intelligent questions.

They will be fluent in the skills such as structured problem-solving and analytical reasoning, evaluating AI responses for bias and accuracy, understanding how prompts shape outcomes, and ethical reasoning and evidence-based decision-making. Setting them up for future jobs such as AI Prompt Engineer (crafting precise questions for intelligent systems); Innovation Consultant (combining human insight with AI analytics); Digital Research Specialist (verifying and refining AI-assisted data); Ethics Officer (ensuring responsible, fair use of technology)

✏️ Marketing Skills and Strategies

Marketing is no longer just about selling products anymore, it's about communicating values, authenticity, and meaning in a digital world saturated with information. Teaching children marketing skills helps them understand influence, creativity, and ethical persuasion. They learn to combine emotional intelligence with strategic thinking which is the ultimate human advantage. In the age of AI-driven advertising and content generation, people who can humanise technology and connect with real emotion will lead the way.

They will be fluent in skills such as human-AI collaboration in creative storytelling, data interpretation and consumer empathy, ethical marketing and responsible communication and digital branding and personal storytelling. Their job prospects include interesting careers such as Digital Marketing Strategist (combining analytics with human creativity),Brand Storyteller (crafting authentic narratives for global audiences), AI-Ethics Marketer (ensuring campaigns remain honest and inclusive) or Social Entrepreneur (building purpose-driven brands in a connected world)

▤ Practical Technology Skills

From typing fluently to managing files and understanding cybersecurity, these skills form the practical toolkit of a modern learner. They give children confidence to navigate, troubleshoot, and create in a technology-first society. They also nurture responsibility knowing how to protect data, manage digital reputation, and use AI tools safely. The ability to adapt to new systems, maintain digital hygiene, and verify information ensures your child stays capable and safe in any AI environment.

The practical skills which are needed include digital literacy and organisation, cyber-safety, password management, and data privacy as well as problem-solving through technology and systems thinking and finally cloud collaboration and productivity tools. These skills support jobs in AI such as Data Security Analyst (protecting online systems from digital threats), UX/UI Designer (improving the interface between humans and AI), IT Support Engineer (troubleshooting human-machine systems) or Digital Project Manager (leading teams in hybrid AI workspaces).

Final Thoughts

The Advanced Skills stage represents the point where learning meets leadership. These are the skills that turn young people from users into creators, from curious learners into innovators. By practising prompting, ethical marketing, and responsible tech use, your child is preparing to thrive in careers that don't yet exist but will soon define our world: human-AI collaboration, digital innovation, and value-driven leadership.

ADVANCED SKILLS AT A GLANCE

PROMPTING CRITICAL THINKING

AI may hold endless data, but it still depends on the clarity and curiosity of the human asking the question. Preempting and critical thinking teach children to communicate precisely, evaluate results and separate usful information from noise. They learn to guide AI systems and not to simply accept their answers, in future, success will belong to those who can think creatively, challenge assumptions and design intelligent questions.

- Structured problem solving and analytical reasoning.
- Evaluating AI responces for bias and reasoning.
- Understanding how prompts shape outcomes

AI prompt Engineer
Innovation consultant

Digital Research Specialist
Ethics Officer

MARKETING SKILLS AND STRATEGIES

Marketing is no longer just about selling products anymore, its about communicating values, authenticity and meaning in a digital world saturated with information. Teaching children marketing helps them understand influence, creativity and ethical persuasion. They learn to combine emotional intelligence with stratigic thinking - the ultimate human advantage. In the age of AI advertising and content generation, People hwo can humanise technology and connect with real emotion will lead the way.

- Human AI collaboration in creative storytelling.
- Data interpretation and consumer empathy.
- Ethical marketing and communication.

Digital Marketing Strategist
Brand Storyteller

AI-Ethics Marketer
Social Entrepreneur

PRACTICAL TECHNOLOGY SKILLS

From typing fluently to managing files and understanding cybersecurity, these skills form the practical toolkit of a modern learner. They give children the confidence to navigate, troubleshoot and create in a technology-first society. They also nurture responsibility knowing how to protect data, manage digital reputation and use AI tools safely. The ability to adapt to new systems, matain digital hygine and verify information ensures your child stays capable and safe in any AI enviroment.

- Digital and technical literacy
- Problem-solving and systems thinking
- Collaboration and project skills

UX/UI Designer
Data Security Analyst

IT Support Engineer
Digital Project Manager

6

PRACTICAL WAYS TO UP-SKILL YOUR CHILD.

We have talked about so many ways in which we can up-skill our children. We have covered;

- How to help children have a good level of emotional intelligence.
- Creating resilient and adaptable people.
- Teaching children how to protect themselves, and their data.
- Helping children understand how to verify data and learn research techniques.
- Helping children to understand the process of problem solving any situation or problem.
- How to prompt AI to do what they want.
- Practical Technology Skills.

These skills are important, however we should always remember that we are **moulding new young minds.** They learn through repetition which is key to long term memory and success.

Soft skills are becoming the most essential part of growing up in this AI world. It is where humans and technology differ. The soft skills include things such as; manners, kindness, gratitude, control of their emotions, reading the room, and reading body language just to name just a few. The earlier you reinforce these soft skills, the easier it is for them to do it without thinking. The last set of skills which are required for GenAI are life skills. By teaching life skills they are going to need as adults, we are **setting them up for success.**

Making sure that they physically experience how to do house chores, gardening and general house maintenance can really develop many of the other skills they need in the AI world coming like problem solving, critical thinking and learning how to assess a situation and make a plan. By introducing all of this in the early years, you have room for lots of repetition. Many people would have heard of the saying "Practise makes perfect" however I love to use a different phrase.

"Practice makes permanency"

This phrase perfectly describes what happens when we use repetition. Whatever you repeat, you will make that skill permanent. It get's hard-wired in the brain.The benefit of this is that you can do that skill and have brain power left to think and wonder.

Have you ever been cleaning and got into the rhythm and starting thinking about other things. The physical actions allow your brain to start wandering to places it would never have been to otherwise. This comes through practising the skill over many years. Working with the brains neuroplasticity, especially when they are young will help prepare them for their future. So it follows that the more you ask your children to do a cleaning job in the house, the better it is for them in the long run. It is hard-wiring that skill in the brain so that they can eventually think more creatively and innovatively while doing that chore later in life.

"Chores are good to learn from 2 years of age."

I was always made to do the bathroom and cleaning when I was a kid. No questions asked, every week I washed and cleaned for 2 hours on a Saturday and then we would go shopping for the weekly dinners. It was never a chore because Dad made it lots of fun. We would chat and talk about life, the universe and everything in between. I remember those shopping trips fondly and still find it fun to clean (even though my robot vacuum, washing machine and dishwasher takes care of the hard labour). I just don't even think about it. Some of my friends still complain about how they have to cook each night, empty the dishwasher and clean, but I just do it as it takes no brain power. Through repetition it has become **'hard wired'** in my brain and I often use those times to think about other issues or projects I am working on. Having my hands busy frees up my brain to become creative and find new innovating ways to think.

Fast forward to my children who were asked to do limited chores. I see that the chores I made them do repetitively when they were little, are done in their houses without a second thought and enjoyed. The chores I chose to always do for them are often left and complained about as being too hard. The hard ones use lots of brain and emotional power. This is your invitation to get your children working in the house and being responsible for doing chores. It will make their lives so much easier when they are adults. Don't advocate to pay them. They are part of the family and it is a family responsibility. Jobs are normally changed as they grow. If you are fortunate enough to have a dishwasher or robot vacuum, get them to organise the vacuum robot to clean twice a week and make sure everything is clear for him to do his job. Learning to delegate is a great skill to have and the more they do this, the easier to becomes later in life. It will teach them how to take responsibility for others work which is an important skill for leadership positions. It is all about teaching responsibility, collaboration with AI and how to work smarter, not harder. So, lighten your load and get your children involved.

7

WHERE TO FROM HERE?

Just as my father was once criticized for allowing his academically gifted child to learn typing in Year 8 and 9, you too may face criticism for the parenting choices you make. Be confident in your decisions. You are supporting your child's growth and helping them develop the essential skills they'll need for the future.

Giving young children time to grow and learn without technology before the age of five is a gift you are giving them no matter how much they think differently. It allows their brains to develop the foundational skills needed for lifelong learning. There will be plenty of time later for screens and technology, but by then, your child will not only know how to use tech, they'll know how to think critically, explore, and innovate with it. If you've already introduced screens, don't worry. It's never too late to start making small changes. And you don't have to do it all at once. This book is filled with ideas, but you only need to start with one. Try it. See how it works for your family.

The most powerful gift you can give your child is a love of learning and a curiosity about the world. That's what builds lifelong learners and in today's fast-changing world, that's more important than ever. You'll discover that most of what your child learns comes from simply being with you, listening to your stories, watching how you live, and absorbing the values you model every day.

What you say becomes their inner voice. So let it be positive.

Children learn through play, observation, and interaction. They even learn from boredom. In fact:

Boredom fuels creativity.

Give them space to feel bored. That's when they learn to think independently, solve problems, and come up with new ideas. These are the kinds of skills that no machine can replace.

Yes, AI is changing the world. Some jobs will disappear, but many more will be created. What won't change is the need for human connection, empathy, collaboration and creativity. These are the skills that set us apart from the AI machines, and these are the skills your child will develop through the experiences, conversations, and shared moments you offer them now.

AI may shape the future, but it cannot speed up how a child's brain grows.

So be present. Play. Explore. Enjoy this time. It goes by so quickly. Remember that even everyday chores can be a chance to connect with your children. Children love spending time with their parents and extended family. They love talking, asking questions, and discovering new things even while they are doing chores together. Don't just let them sit there, encourage them to be part of the family at dinner time and when it's time to do chores. If they love their screens, that's okay too. Focus on **balance**, **connection**, and **consistency**.

A calm, structured environment helps their brains feel safe and when they feel safe, they're free to learn.

Above all, remember: **you are their most important teacher**. What you do matters more than you know. You're not just raising a child. You're shaping a future.

Final Thoughts

Thank you for joining me on this journey through *The Advantage Code*. By reading this book, you've taken a powerful step toward equipping your child with the tools, skills, and mindset they'll need to thrive in a world shaped by AI and constant change.

Parenting has never been a simple task, but in today's rapidly evolving landscape, your role is more important than ever. The fact that you've invested the time to explore these strategies shows your deep commitment to helping your child grow into a capable, compassionate, and confident individual.

As you move forward, remember that future readiness isn't about having all the answers — it's about nurturing curiosity, resilience, and a willingness to learn and adapt. No matter where your child's journey leads, the skills you've helped them develop will act as a strong foundation.

Thank you for trusting me to be a part of your parenting journey. I hope the ideas and strategies in this book inspire you to keep growing alongside your child and to create a home filled with love, learning, and limitless possibility.

The future is coming quickly — but with your guidance, your child will be ready to embrace it with courage and joy.

"The best way to predict the future is to prepare our children to shape it."

8

SUPPORTING INFORMATION AND TIPS

Books

Biddulph, Steve: Raising Boys, Raising Girls

Covey, Sean: 7 Habits of Highly Effective Teens, 7 Habits of Highly Effective People

Duckworthy, Angela (2007): Grit The power of passion and perseverance, Why passion and resilience are the secrets to success

Gladwell, Malcolm: Outliers, Tipping Point

Green, Dr Christopher: New Toddler Taming

Siegel, Daniel: The Whole-Brain Child

Research Papers

Arain, M., Haque, M., Johal, L., Mathur, P., Nel, W., Rais, A., Sandhu, R., & Sharma, S. (2013). Maturation of the adolescent brain. *Neuropsychiatric disease and treatment, 9*, 449–461. https://doi.org/10.2147/NDT.S39776

Carnegie, D. (1936). How to win friends and influence people. *Simon & Schuster*

Domingues-Montanari, S. (2017) Clinical and psychological effect of excessive screen time on children. *Journal of Paediatrics and Child Health.* Https://doi.org/10.1111/jpc.13462

Du F. From Playground to Boardroom: Endowed Social Status and Managerial Performance. *Journal of Financial and Quantitative Analysis.* 2022;57(3):988-1022. Dot:10.1017/s0022109021000089

Gest, S. D., Sesma, A., Masten, A. S., & Tellegen, A. (2006). Childhood peer reputation as a predictor of competence and symptoms 10 years later. Journal of Abnormal Child Psychology, 34, 509–526.

Muppalla S, Vuppalapati S, Reddy Pulliahgaru A, et al. (June 18, 2023) Effects of Excessive Screen Time on Child Development: An Updated Review and Strategies for Management. Cureus 15(6): e40608. DOI 10.7759/cureus.40608

Obschonka, Silbereisen, & Schmitt-Rodermund (2011). *Successful Entrepreneurship as Developmental Outcome: A Path Model from a Lifespan Perspective of Human Development.* European Psychologist 16(3):174-186. DOI:10.1027/1016-9040/a000075

Z. A. Ranganathan, "How Much Screen Time Should children Have?," *2024 IEEE Integrated STEM Education Conference (ISEC)*, Princeton, NJ, USA, 2024, pp. 1-1, doi: 10.1109/ISEC61299.2024.10665189.

Robinson R (2009) From Child to Young Adult, the Brain Changes Its Connections. PLoS Biol 7(7): e1000158. doi:10.1371/journal.pbio.1000158

Supekar, K, Musen, M., Menon, V., (2009). Development of Large-Scale Functional Brain Networks in Children. PLOS Biology. https://doi.org/10.1371/journal.pbio.1000157

PIAGET STAGES OF COGNITIVE DEVELOPMENT	
Sensorimotor Stage (0-2 years)	Infants learn about the world through their senses and motor skills, using their hands and mouth. Their language is used primarily for demands and cataloguing. They recognise their own self as an agent of action and acts intentionally (e.g. pull a toy to make it move or a noise). Children acts on the here and now and are very egocentric. Milestone - They develop object permanence during this time, which is the understanding that objects continue to exist even when out of sight. They know that Mum or Dad exist even when they aren't around.
Pre-operational Stage (2-7 years)	Children use symbolic thought such as language and imagination to represent objects and events. They can classify objects by a single feature eg. colour, size, shape. Children talk about and understand things beyond their immediate experience. Limitations - Children's thinking is still egocentric and can't see others perspectives, they lack logical operations and think more in terms of words and symbols
Concrete Operational Stage (7-11 years)	Children develop logical thinking about concrete events and tangible objects- They understand Math concepts such as time, space, and quantity and can be applied but not as independent concepts. Milestones - They gain an understanding of conservation which means that quantity remains the same despite changes in appearance, reversibility which means that actions can be reversed. children organise and classify objects as well as understanding the difference between truth and lies.
Formal Operational Stage (12+ years)	Adolescents and adults develop the ability to think abstractly and hypothetically.They can use critical thinking and problem solving techniques in their everyday life. Milestones - Children use deductive reasoning, think about complex problems and understand abstract concepts like justice and morality.

Developmental Stages 0 - 5 years

This table is a collation of information from many sources. It contains general information. If you are worried about your child's development please seek professional medical advice as all children develop at different rates and they will be able to set your mind at rest or know what action to take for early intervention.

AGE	MOTOR	SPEECH AND VOICE	VISION AND HEARING	SOCIAL DEVELOPMENT	MUSICAL ACTIVITIES TO HELP THEIR DEVELOPMENT
0-2mths	Reach for objects	vocalises	15cm from face	Smiles, laughs	Sing to your child
2 - 6 mths	Starts to be able to control muscles, eg starts to feed themselves, rollover, sit, maybe crawl. • Action songs • Beat on body	Say mama, dada, Uses voice more to communicate (ie. Different cries for specific needs, I'm hungry, I'm bored. Imitates coughs or other sounds. Uses different pitch and experiments with dynamics (loud,soft)	Localises sound 45cm lateral to either ear. Shows a preference to music and sights Sensitive to sound and vision (ie. Too bright, or too loud) • Singing • Soft music • silence	Squeals with delight. Smiles and laughs to your smiles. Smiles at their mirror image. May show 'stranger shyness' • action rhymes /songs	Shakers to music Listening to a variety of music (they will have a preference which may change as they grow) Sing nursery rhymes. Read books (Rhyming books or singing books are best) Times of silence during the day.
6-9 mths	Rolls over, can sit up, crawls and stands supported. Starts to gesture for objects or things (eg. Read a book, pointing to that apple they want to eat.) Starts to move more freely. Can move to objects they want	More babbling like words. Uses pitch and dynamics of sounds more • increase vocalisation range with sounds around them (eg animals)	Has real preferences to types of music and knows how to display their likes and dislikes • play variety of music • be guided by child • jazz, regae, classical, baroque	Respond to their name. Responds to telephone, or knock on door, Understands "no" Starts to PLAY Enjoys games such as peek-a-boo, tickling etc. Enjoys repetitive books, and the same book repetitively.	Variety of classical, country, jazz, and nursery rhymes played everyday. Times of silence. Learn to listen to music around them (eg birds singing, trees rustling) Starts to "sing" nursery rhymes or part of them.

AGE	MOTOR	SPEECH AND VOICE	VISION AND HEARING	SOCIAL DEVELOPMENT	MUSICAL ACTIVITIES TO HELP THEIR DEVELOPMENT
9-18 mths	Starts to walk. Holds toys and moves them where they want. Starts to climb • Dance • Hits drums • shakes	Combines sounds to start talking Uses combination of words to get things Uses dynamics during speech and tries to sing favourite songs Can hear phrases of songs before words are clear. • singing nursery rhymes • scarf vocalising • reading rhyming books	Loves listening to rhyming sounds (eg Cat in the hat style of books) Still sensitive to sounds. Enjoys variety of music. Follows objects which are dropped. • Singing live • Silence • Playing variety of music • Repetition of music liked • Repetition of books/rhyming	Performs for attention Shows interest in picture books, loves repetition. Follows simple directions (eg. sit down) Understands words Recognises favourite songs. Finds objects which are dropped. Understands "in and out", "off and on". Knows some body parts Starts to do pretend play. Points to pictures • Reading books • Singing live • Action songs	Shakers to music Drum to the beat of the music on a drum. eg. Tupperware upside down with a spoon/ stick/etc. Pots and pans, buckets upside down, Anything that makes a cool sound. Bells Clapping or tapping to music Dancing to music Chimes (spoons, forks, metal objects hanging down and hit with a metal spoon) Singing songs with animals in them (eg. 5 little ducts) and body parts (eg. Heads shoulders)

AGE	MOTOR	SPEECH AND VOICE	VISION AND HEARING	SOCIAL DEVELOPMENT	MUSICAL ACTIVITIES TO HELP THEIR DEVELOPMENT
18mths – 3 years	Starts to run and never stops. Can climb stairs Climbs anything (tables, play equipment, chairs) Turn pages of books. Jumping and kicking skills are developed during this period. • Lots of action songs • Play the beat on the drums • Play the beat on wood blocks • Making music with home made instruments together	Uses action works Produces words with two or more syllables Starts talking in sentences. Can speak more than 350 words. Hold a simple tune (eg incy wincy spider) Talks about past events. Can be understood by anyone. • Can sing more complex songs • Try rhyming poems to known tunes • Repetition • Sequencing songs – what comes next	They can pick out sounds clearly. Imitate most sounds. Have discovered a rhythm to speech patterns. Aware of the function of print. Recognizes signs and logos Can recognize and name colours, shapes and objects. Distinguish between different colours of the same type (eg blue, aqua, navy) • can hear harmonies • sing in rounds • read, read • sing songs which they are interested	Follows directions. Understands concepts such as you, me, mine. Takes turns Shows concern for others Imaginative play Combines several actions in play Understands concepts of size (big/little) and quantity (a little, a lot or more) Understands, who, what, where and why. Loves craft. • Make own instruments • Sing parts of stories	Making instruments like shakers and drums to play with. Listening to a wider range of music. Listening to one song repetitively. Sings familiar songs (eg nursery rhymes) Using familiar objects to make their own music. Dancing to familiar music Keep the beat on wood blocks or drum. Loves the Wiggles. Nursery rhymes with large body movements (eg, ring a ring a rosie, Punchinello, Mulberry bush Sing acting out nursery rhymes (eg Pussy cat, Mr frog, 5 little ducks)

AGE	MOTOR	SPEECH AND VOICE	VISION AND HEARING	SOCIAL DEVELOPMENT	MUSICAL ACTIVITIES TO HELP THEIR DEVELOPMENT
3-4 years	Increases skills in large motor movement such as running, climbing, jumping, and kicking. Learns to hold a pencil Cut with scissors Glue and paint Starts to perfect small motor movements such as colouring in, letter formation, cutting with scissors, glueing, painting etc. • Makes own instruments. • More complex rhythms on the drum • Able to start an instrument	Uses adult grammar. Uses more complex words such as; before/after between/beside same/different rough/smooth and easy/difficult. Describes events in detail. Tells stories with a beginning, middle and end. Starts to find rhyming words, eg cat-hat-bat etc. Matches some letters to sounds Describe a story or movie in their own words. Makes up poems and songs.	Can distinguish a wider variety of sounds. Understands high and low notes, fast and slow music. Enjoys particular type of music. Sing all songs in tune. Use to all the sounds of their language • sing in rounds • sing different parts • make own songs with words	Follows directions involving 3 or more steps. Develops more complex imaginative play. Starts to solve problems by working through the steps. Starts to understand the sequencing of events. Loves repetition Constantly asks questions. Learns to dress and undress. Understands directions with "if and then" in them. Seeks to please his/her friends Independence in friendships Starts to help with tasks.	Making a washboard. Experiments with different instruments to make different sounds, eg water xylophone. Chimes of different notes to make musical songs. Making up words to old songs. Humorous songs - Mr Clickety Cane, Dur glump, Mr Goat. Action songs - a sailor went to sea Start an instrument, violin, piano, viola, cello, flute, and guitar. Sing in rounds. Beat the rhythm or beat of songs and rhymes. Drum different complex rhythms. Makes their own compositions